Godwin Study Guides

Structural Steelwork

Third Edition

F. W. Lambert

GODWIN STUDY GUIDES

STRUCTURAL STEELWORK

Godwin Study Guides

Materials and Structures
Theory of Structures
Advanced Theory of Structures
Design of Reinforced Concrete Elements
Specifications and Quantities
Surveying
Soil Mechanics
Fluid Mechanics
Mathematics for Engineers

GODWIN STUDY GUIDES

STRUCTURAL STEELWORK

F. W. LAMBERT
M.Phil., C.Eng., M.I.C.E., M.I.Struct.E.
Senior Lecturer in Structural Engineering
Polytechnic of the South Bank

THIRD EDITION

GEORGE GODWIN
LONDON

George Godwin,
Longman House,
Burnt Mill, Harlow, Essex, UK

A division of Longman Group Ltd, London

Published in the United States of America
by Longman Inc., New York

First published 1973 by Macdonald & Evans Ltd
in the *Examination Subjects for Engineers and Builders* series

Second edition 1977

This edition published 1982 by George Godwin as a *Godwin Study Guide*

British Library Cataloguing in Publication Data

Lambert, F. W.
  Structural steelwork – 3rd ed.
  1. Steel, Structural
  2. Structural design
  I. Title
624.1′821   TA684

ISBN 0–7114–5712–3

Printed in Great Britain by
Richard Clay (The Chaucer Press) Ltd, Bungay, Suffolk

GENERAL INTRODUCTION

THIS series was originally designed as an aid to students studying for technical examinations, the aim of each book being to provide a clear concise guide to the *basic principles* of the subject, reinforced by worked examples carefully selected to illustrate the text. The success of the series with students has justified the original aim, but it became apparent that qualified professional engineers in mid-career were finding the books useful.

In recognition of this need, the books in the series have been enlarged to cover a wider range of topics, whilst maintaining the concise form of presentation.

It is our belief that this increase in content should help students to see their study material in a more practical context without detracting from the value of the book as an aid to passing examinations. Equally, it is believed that the additional material will present a more complete picture to professional engineers of topics which they have not had occasion to use since completing their original studies.

A list of other books in the series is given at the front of this book. Further details may be obtained from the publishers.

M. J. Smith
General Editor

AUTHOR'S PREFACE

THIS book is about structural steelwork and is concerned with an important part of the total design process, namely, the detailed design of the elements of structure. Viewed in the broader sense, structural design embraces the planning of the framing system, estimating the loading and calculating the resulting forces and moments, designing the individual members and the preparation of the details for construction.

Modern structural steelwork relies heavily on welding and bolting and this is reflected throughout the text. Riveted structures are now virtually obsolete and for this reason will not be discussed.

The first four chapters deal with the design of simple elements of structure complying with the general principles of B.S. 449 *The Use of Structural Steelwork in Building*. Chapter 5 covers the plastic theory methods of design for which there is at present no British Standard, although this subject is now well documented, and chapter 6 deals with composite design techniques. Both chapters 5 and 6 reflect a more efficient approach to element design, than that produced by simple elastic theory methods.

It is assumed that the reader is familiar with elementary theory of structures and strength of materials which are essential tools in the design process.

In producing the third edition of *Structural Steelwork* the opportunity has been taken of including additional worked examples and revising, where appropriate, parts of the existing text. The chapter on trusswork and bracing is completely new. Worked examples have been prepared in S.I. units, using a pocket calculator and the values rounded up for practical purposes.

August 1981   F.W.L.

CONTENTS

LIST OF SYMBOLS

p	Permissible stress (subscript denotes type)
f	Working or actual stress
f_y	Yield stress
f_{cu}	28-day concrete cube strength
q	Shear stress (plastic design)
c_s	Critical stress
$\left.\begin{array}{c} F \\ P \end{array}\right\}$	Force—permissible force (subscript denotes type)
Q	Shear force
W	Load or force as appropriate
R	Reaction
M	Bending moment
M_u	Ultimate moment (composite design)
M_p	Plastic moment
I	Moment of inertia
Z	Section modulus (elastic design)
Z_p	Plastic modulus
m	Shear centre ratio
r	Radius of gyration
$\bar{Y}$	Neutral axis distance
A	Area
L	Length or height
h	Height
D	Depth
B	Breadth
l	Effective length
d	Depth, diameter—spacing
b	Breadth
T	Thickness—flange
t	Thickness—web, etc.
e	Eccentricity
q	Effective length factor
Kk	Modification factor (subscript denotes type)
β	factor in plastic design
φ	angle
N	Number of bolts, shear connectors
n	Centroidal distance, factor, number
$\left.\begin{array}{c} x \\ y \\ a \end{array}\right\}$	Small distances
s	Spacing, weld size

CHAPTER 1

CONNECTIONS

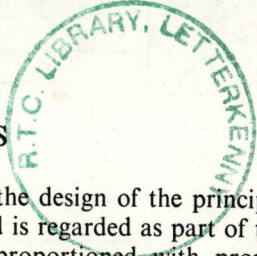

THE design of connections usually follows the design of the principal components of a steel-framed structure, and is regarded as part of the detailing process. Connections must be proportioned with proper regard to the design method adopted for the whole structure.

BOLTED CONNECTIONS

In the structural steelwork three types of bolt are generally available These are:

1. Black bolts—of mild steel quality designated grade 4.6 or higher strength bolts designated grade 8.8 are used where no great accuracy is required and where vibration, stress reversals, fatigue, etc., will not be encountered. They are inserted into holes nominally 2 mm larger than their diameter.

2. Machined bolts—of similar quality to that given above, but used where a higher standard of workmanship is demanded. The shank or barrel is accurately machined and the hole into which they are inserted should not exceed the bolt diameter by more than 0·15 mm.

3. Friction-grip bolts—these are produced from high-grade steel and are used for all important connections. They have replaced rivets and largely superseded machined bolts, and are covered by their own material and design specification.

Bolt sizes must be practical; 16 mm diameter may be regarded as minimum. In any one connection all bolts should be of equal size and types should not be mixed.

Strength of black and machined bolts

The design strength of a bolt is either its shear or bearing capacity, whichever is the lowest.

In single shear, the capacity of one bolt is

$$F_s = \frac{\pi d^2}{4} \cdot p_s,$$

and in bearing

$$F_b = dt \cdot p_b$$

Where bolts are in double shear the value of F_s calculated above may be multiplied by 2. No adjustment should be made to the bearing value F_b. Single and double shear are illustrated in Fig. 1.

1

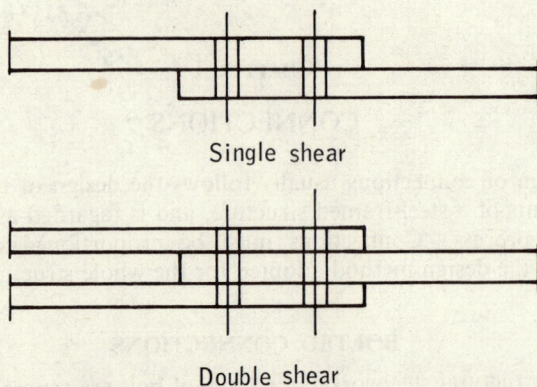

Single shear

Double shear

Fig. 1

For a bolt in tension, the design strength is given by

$$F_t = A_s p_t$$

in which A_s is the tensile stress area as defined in B.S. 4190. Allowable stresses in bolts are given in Table 20 of B.S. 449 amendment AMD 1787.*

Combined shear and tension

Where bolts are resisting both shear and direct tension, the ratio of their actual to permissible stresses should be proportioned such that

$$\frac{f_s}{p_s} \not> 1$$

$$\frac{f_t}{p_t} \not> 1$$

and

$$\frac{f_s}{p_s} + \frac{f_t}{p_t} \not> 1 \cdot 4$$

Shown graphically this means that these three ratios should fall within the shaded area of Fig. 2.

Almost the same result would obtain using the relationship

$$\left(\frac{f_s}{p_s}\right)^2 + \left(\frac{f_t}{p_t}\right)^2 \not> 1$$

Whilst the design specification B.S. 449 gives no explanation for the basis of the adopted approach there is experimental evidence to show

* Unless otherwise stated, all numbered tables in the text refer to those in B.S. 449.

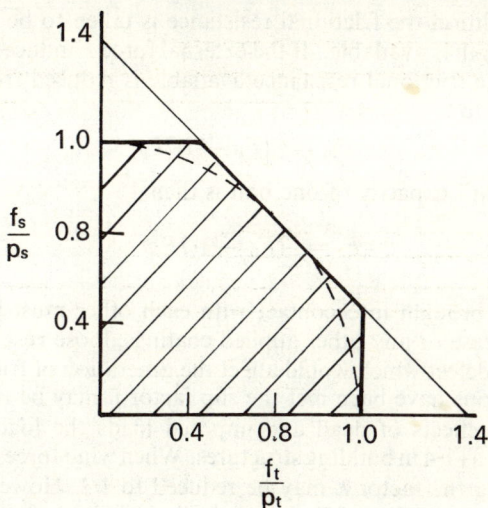

Fig. 2

that this interaction equation (shown as a dotted line in Fig. 2) is representative of both groups subject to combined stresses.

Strength of friction-grip bolts

The design concepts for this type of bolt are quite different from that given above. Referring to Fig. 1, it is assumed that the "shear" capacity is obtained by the resistance to friction induced between the plate interfaces local to the bolt, which is caused by stressing the bolt in tension to its proof load.

If μ is the coefficient of friction and F_p the proof load the maximum frictional resistance of one bolt is

$$F_m = F_p \cdot \mu,$$

and applying a load factor λ against slip and assuming there are n interfaces the "shear" capacity of one bolt will be

$$F_s = F_m \frac{n}{\lambda}$$

or

$$F_s = \frac{\mu}{\lambda} \cdot F_p \cdot n$$

Where these bolts are arranged such that they are subject to direct tension from an externally applied connection force, the design capacity in tension should not exceed

$$F_t = 0 \cdot 6 F_p$$

For this condition the frictional resistance is taken to be zero and no "shear" strength is available. If the external force F induced is less than $0.6F_p$ then the frictional resistance available is reduced from that previously given to

$$F_m = \mu(F_p - 1.7F)$$

and the "shear" capacity of one bolt is then

$$F_s = \frac{\mu}{\lambda}(F_p - 1.7F)n$$

The surfaces brought into contact with each other must be free from paint, oil, grease or any other applied coatings, loose rust or millscale, or any other defect which would affect the generation of friction. Where these conditions have been met the slip factor μ may be taken as 0.45.

Under the effects of dead and imposed loads the load factor λ is usually taken as 1.4 in building structures. When wind forces are included in the loading the factor λ may be reduced to 1.2. However, it is important to ensure that each loading condition is investigated and bolts provided for the worst condition.

Generally plate thickness or plies should not be less than 10 mm thick or half the bolt diameter whichever is the less.

Design of bolted connections

Three rules apply to the design of bolted connections:

1. the product of the number of bolts used and their strengths must not be less than the force they are required to resist;
2. the moment of resistance of a bolt group must not be less than the moment they are required to resist;
3. the plates and/or sections employed to form the connection of which the bolts form part must be capable of resisting safely the forces and moments to which they are subject.

EXAMPLE 1

A simple tension member of grade 43 steel is fabricated from a 75 mm × 15 mm flat, and the force resisted is 123 kN. Determine the number of black bolts required assuming they are 20 mm diameter and operate in double shear. What size splice plates are required to transfer this force across the joint in the main member?

SOLUTION

From Table 20

$$p_s = 80 \text{ N/mm}^2 \qquad p_b = 250 \text{ N/mm}^2$$

In double shear, capacity of one bolt is

$$F_s = 20^2 \times \frac{\pi}{4} \times 80 \times 2 \times 10^{-3} = 50.3 \text{ kN}$$

In bearing, capacity of one bolt is

$$F_b = 20 \times 15 \times 250 \times 10^{-3} = 75 \text{ kN}$$

Since double-shear capacity governs, the number of bolts required is

$$N = \frac{W}{F_s} = \frac{123}{50 \cdot 3} = 2 \cdot 45, \text{ say 3.}$$

From Table 19

$$p_t = 155 \text{ N/mm}^2$$

As joint is the double-shear type, force in each splice plate is clearly 50 per cent of 123 kN, *i.e.* 61·5 kN and the net required for each plate is therefore

$$A_n = \frac{W}{p_t} = \frac{61 \cdot 5 \times 10^3}{155} = 397 \text{ mm}^2$$

Making the plate width 75 mm, less one hole 22 mm diameter, the net thickness is

$$T(75 - 22) = 397$$
$$\therefore T = 7 \cdot 5, \text{ say 8 mm.}$$

This connection detail is shown in Fig. 3.

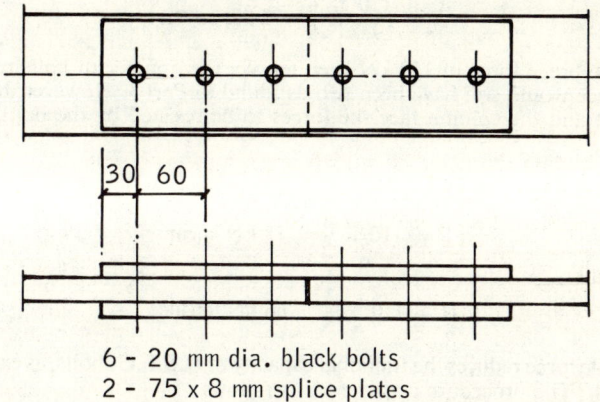

6 - 20 mm dia. black bolts
2 - 75 x 8 mm splice plates

FIG. 3

EXAMPLE 2

A braced bay in a building is arranged geometrically as a 3-4-5 triangle. The diagonal brace consists of a pair of 100 × 75 × 8 angles, longer leg connected, and carries an axial compressive force of 110 kN. Design the terminal connections using 16 mm diameter friction-grip bolts, and assume the terminal gussets are symmetrically connected to the flange of a 203 × 203 × 46 U.C.

SOLUTION

This problem is in three parts:

1. design of bolts between brace and gusset;

2. design of bolts between gusset and U.C.;

3. design of gusset.

From B.S. 4604, $\mu = 0.45$, $\lambda = 1.4$, and Table 2 gives $F_p = 92.1$ kN.

Part 1

Gusset will pass between braces giving a "double-shear" effect

$$\therefore \; F_s = \frac{0.45 \times 92.1 \times 2}{1.4} = 59.2 \text{ kN per bolt}$$

and the number of bolts required will be

$$N = \frac{110}{59.2} = 1.86, \text{ say } 2.$$

Part 2

The gusset flange is connected to the flange of the U.C. and operates in "single shear." Only the vertical component of the brace force need be considered as the horizontal component produces compression between the bolt interfaces, then

$$F_s = \frac{0.45 \times 92.1}{1.4} = 29.6 \text{ kN per bolt}$$

and the number of bolts required will be

$$N = \frac{110}{29.6} \times \frac{4}{5} = 2.97, \text{ say } 4.$$

Had the brace force of 110 kN been tensile, the number of bolts required in the brace would still have been two as found in Part 1. However, between the gusset and the column face the forces to be resisted by the bolts would be:

Vertically

$$V = 110 \times \frac{4}{5} = 88 \text{ kN shear}$$

Horizontally

$$H = 110 \times \frac{3}{5} = 66 \text{ kN tensile}$$

This tensile force reduces the frictional capacity of H.S.F.G. bolts, as explained on page 4. The procedure is therefore as follows:

Try four 16 mm H.S.F.G. bolts

External tension $= \dfrac{66}{4} = 16.5$ kN each

therefore, net frictional resistance of one bolt is

$$F_s = \frac{0.45}{1.4} (92.1 - 1.7 \times 16.5) = 20.6 \text{ kN}$$

and for four bolts $V = 4 \times 20.6 = 82.5$ kN

Since this is less than the applied shear of 88 kN the detail is inadequate. Therefore, the number of bolts should be increased.

Try six, giving an external tension of 11 kN per bolt.

For six bolts $V = \dfrac{6 \times 0.45}{1.4} (92.1 - 1.7 \times 11)$

$$= 141 \cdot 6 \text{KN} > 88 \text{kN}$$

This is now more than adequate and whilst five bolts can be shown to be sufficient the detail would not be very practical.

Part 3

Use a beam cutting for the gusset Ex. 406 × 178 × 74 U.B. and shape to suit. The web of this beam is 9·7 mm thick and will pass between the braces. Along the bolt line the distance between the face of the U.C. and the first bolt in the brace is about 120 mm and is unsupported in compression. The proportions are checked as follows:

$$r = \frac{t_w}{\sqrt{12}} = \frac{9 \cdot 7}{\sqrt{12}} = 2 \cdot 8 \text{ mm}$$

$$\frac{l}{r} = \frac{1 \cdot 5 \times 120}{2 \cdot 8} = 64 \text{ say}$$

which gives p_c from Table 17a of 122 N/mm². Therefore minimum width of gusset at right angles to bolt line is

$$b = \frac{W}{p_c \times t_w} = \frac{110 \times 10^3}{122 \times 9 \cdot 7} = 93 \text{ mm.}$$

Along the vertical plane in front of the gusset flange the length required to resist shear is, assuming $p_q = 100$ N/mm²,

$$l = \frac{110 \times 10^3}{110 \times 9 \cdot 7} \times \frac{4}{5} = 91 \text{ mm approx.}$$

This connection detail is shown in Fig. 4.

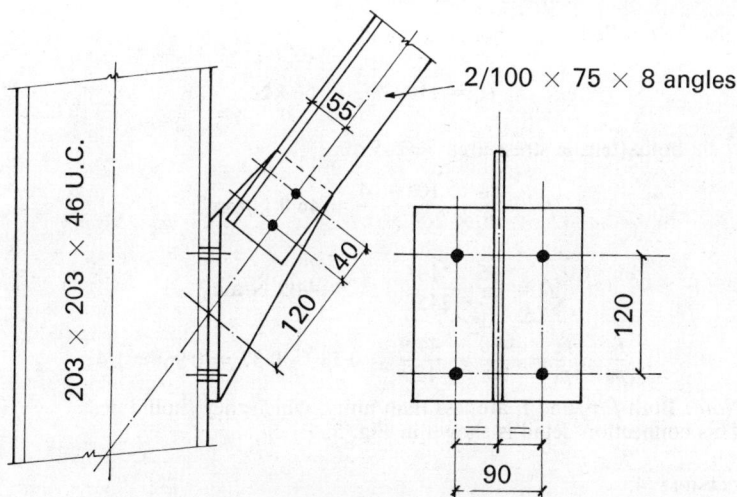

16 mm dia. H.S.F.G. bolts
Gusset Ex. 406 × 178 × 74 U.B.

FIG. 4

EXAMPLE 3

Redesign the bracing connection in Example 2 if the force is 110 kN tension. Use 20 mm diameter black bolts.

SOLUTION

This problem is in two parts:

1. bolts in brace;
2. bolts between gusset and column.

From Table 20

$$p_s = 80 \text{ N/mm}^2,\, p_t = 120 \text{ N/mm}^2,\, p_b = 250 \text{ N/mm}^2$$

Part 1

$$F_s = 20^2 \times \frac{\pi}{4} \times 80 \times 2 \times 10^{-3} = 50 \cdot 3 \text{ kN per bolt,}$$
$$F_b = 20 \times 9 \cdot 7 \times 250 \times 10^{-3} \quad = 48 \cdot 5 \text{ kN per bolt.}$$

Since bearing governs, the number of bolts required will be,

$$N = \frac{110}{48 \cdot 5} = 2 \cdot 26, \text{ say 3.}$$

Part 2

As the brace is in tension, these bolts will be subject to both shear and tension.

The shear force is vertical and is

$$V = 110 \times \frac{4}{5} = 88 \text{ kN,}$$

whilst the tensile force is horizontal and is

$$H = 110 \times \frac{3}{5} = 66 \text{ kN}$$

Try six bolts (tensile stress area = 245 mm²)

$$f_s = \frac{88 \times 10^3 \times 4}{6 \times 20^2 \times \pi} = 46 \cdot 8 \text{ N/mm}^2$$

$$f_t = \frac{66 \times 10^3}{6 \times 245} \quad = 44 \cdot 9 \text{ N/mm}^2$$

$$\therefore \frac{f_s}{p_s} + \frac{f_t}{p_t} = \frac{46 \cdot 8}{80} + \frac{44 \cdot 9}{120} = 0 \cdot 58 + 0 \cdot 37 = 0 \cdot 96 < 1 \cdot 4$$

Note: Both f_s/p_s and f_t are less than unity, which they should be. This connection detail is shown in Fig. 5.

EXAMPLE 4.

The main tie of a heavy-duty lattice girder consists of two grade 50 432 × 102 × 65·54 channels spaced 200 mm back to back, and resists a direct force of 2740 kN. Design completely the splice using grade 50 cover plates and 24 mm diameter friction-grip bolts. In this problem both the flanges and the webs should be spliced to ensure a reasonably distributed stress pattern.

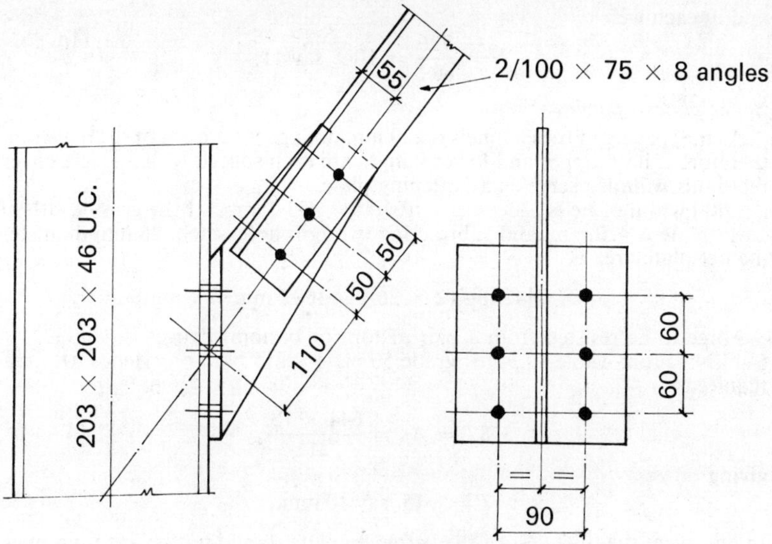

2/100 × 75 × 8 angles

20 mm dia. Black bolts
Gusset Ex. 406 × 178 × 74 U.B.

FIG. 5

SOLUTION

For one channel gross area = 8349 mm²; gross area of web measured between fillets is

$$d_w t_w = 362 \cdot 5 \times 12 \cdot 2 = 4423 \text{ mm}^2,$$

giving a gross area of each flange of

$$A_f = \frac{8349 - 4423}{2} = 1963 \text{ mm}^2$$

Since the force is axial each channel carries 1370 kN. Then force on web is

$$F_w = 1370 \times \frac{4423}{8349} = 726 \text{ kN},$$

and on each flange

$$F_f = 1370 \times \frac{1963}{8349} = 322 \text{ kN}$$

Bolt quantities—from B.S. 4604—$F_p = 207$ kN. The "single-shear" value will be

$$F_s = \frac{0 \cdot 45 \times 207}{1 \cdot 4} = 66 \cdot 5 \text{ kN}$$

The number of bolts required in each flange is

$$N = \frac{322}{66 \cdot 5} = 4 \cdot 84, \text{ say } 5$$

and in each web

$$N = \frac{726}{66 \cdot 5} = 10 \cdot 9, \text{ say } 11.$$

Splice plates—flanges

As the spacing of the channels is 220 mm, web packs are out of the question, therefore if both upper and lower flanges are each spliced with a single cover plate this will also serve as a battening plate.

Total width of tie is $220 + (2 \times 101 \cdot 6) = 423 \cdot 2$ mm. If the gross width of cover plate is 420 mm, and subtracting two bolt holes each, 26 mm diameter the net plate area is

$$T[420 - (2 \times 26)] = 368T \text{ mm}^2$$

Force to be resisted from a pair of top (or bottom) flanges is $2 \times 322 = 644$ kN. From Table 19 p_t for grade 50 plate is 215 N/mm². Hence net area required is

$$368T = A_n = \frac{644 \times 10^3}{215},$$

giving

$$T = 8 \cdot 15, \text{ say } 10 \text{ mm.}$$

In any event the thickness of the battening plate should not be less than one-fiftieth of the distance between longitudinal bolt lines. This requirement will be seen to be satisfied.

Splice plates—web

A single cover plate was assumed for the number of bolts needed. Bolt this plate on the outside of the web, *i.e.* between flanges. Make net width slightly less than d_w, say 360 mm, and deducting for three bolt holes, each 26 mm diameter, the net plate area is

$$T[360 - (3 \times 26)] = 282T \text{ mm}^2$$

Using $p_t = 215$ as before, net area required is

$$282T = A_n = \frac{726 \times 10^3}{215}$$

giving

$$T = 11 \cdot 95, \text{ say } 12 \text{ mm.}$$

The arrangement of this connection is shown in Fig. 6.

EXAMPLE 5.

A $305 \times 305 \times 118$ U.C. is to be spliced to a $356 \times 368 \times 129$ U.C. The axial load to be transferred from the small U.C. is 800 kN dead load and 1080 kN live load; there is also a major axis bending moment at the splice of 70 kN m. Determine the proportions of a suitable splicing arrangement assuming all material is grade 43.

SOLUTION

For this detail three solutions are possible. If the combined effect of the dead load and moment does not give tension the splice material need only be sufficient to hold the members in position. Should tension develop both splice plates and bolting must be adequate to resist it. Alternatively the entire force system may be resisted by the splice material, which of course is unnecessarily wasteful but sometimes cannot be avoided.

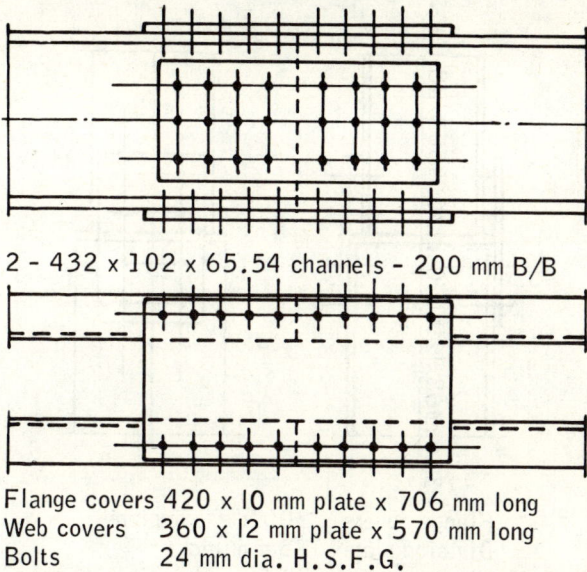

2 - 432 x 102 x 65.54 channels - 200 mm B/B

Flange covers 420 x 10 mm plate x 706 mm long
Web covers   360 x 12 mm plate x 570 mm long
Bolts   24 mm dia. H.S.F.G.

Fig. 6

Check stresses in 305 × 305 × 118 U.C. to determine approach

$$A = 14\,980 \text{ mm}^2, \qquad Z_{xx} = 1755 \times 10^3 \text{ mm}^3,$$

$$f_c = \frac{W}{A} = \frac{800 \times 10^3}{14\,980} = +53 \cdot 4 \text{ N/mm}^2,$$

$$f_{bc/t} = \frac{M}{Z} = \pm\frac{70 \times 10^6}{1755 \times 10^3} = \pm39 \cdot 9 \text{ N/mm}^2$$

$\therefore$ minimum stress $= f_c - f_{bc} = 53 \cdot 4 - 39 \cdot 9 = +13 \cdot 5 \text{ N/mm}^2$

as no tension develops provide nominal splice.

At the splice a division plate should be inserted between the two column lengths to allow for their size difference, and this should be thick enough (25 mm should suffice) to permit the load to spread at an angle of say 45°. The flange cover plates need only be nominal, say about 10 mm in thickness or approximately half the thickness of the thicker flange, and black bolted into position.

The ends of the columns should be machined square with their axes to give dead bearing. This splice is shown in Fig. 7.

EXAMPLE 6.

Using the column sections from the previous example determine the splice proportions if the dead load is 400 kN, the live load 700 kN and the moment is 200 kNm.

SOLUTION

Checking the stresses in the upper column.

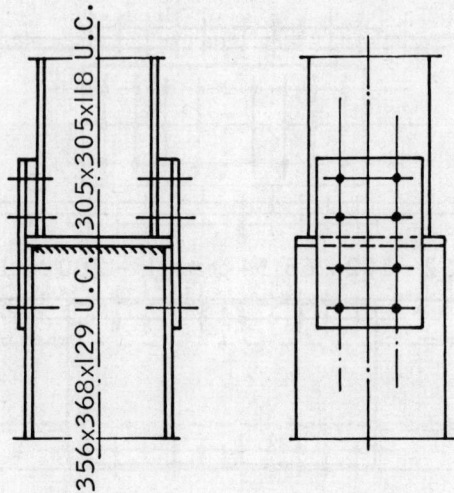

Flange covers   300 x 10 mm plate
Division plate   25 mm thick
Flange packs   thickness to suit
Bolts   22 mm dia. black

Fig. 7

1. Under full loading:

$$f_c = \frac{1100 \times 10^3}{14\,980} = 73 \cdot 5 \text{ N/mm}^2,$$

$$f_{bc/t} = \frac{200 \times 10^6}{1755 \times 10^3} = 113 \cdot 8 \text{ N/mm}^2,$$

$$\therefore \text{ min. stress} = 73 \cdot 5 - 113 \cdot 8 = -40 \cdot 3 \text{ N/mm}^2\text{-tension}$$

2. Under dead loading:

$$f_c = \frac{400 \times 10^3}{14\,980} = 26 \cdot 7 \text{ N/mm}^2,$$

$$f_{bc/t} = \frac{200 \times 10^6}{1755 \times 10^3} = 113 \cdot 8 \text{ N/mm}^2$$

$$\therefore \text{ min. stress} = 26 \cdot 7 - 113 \cdot 8 = -87 \cdot 1 \text{ N/mm}^2\text{-tension}$$

Tension develops in both instances and the worst case for design is dead load plus moment.

Force to be resisted by flange cover plates is

$$F = \frac{W_d}{2} - \frac{M}{D}$$

$$= \frac{400}{2} - \frac{200 \times 10^3}{314 \cdot 5} = -424 \text{ kN-tension}$$

As the number of black bolts required to carry this force will be excessive, use instead 22 mm. diameter H.S.F.G. bolts, $F_p = 177$ kN. For "single-shear" conditions

$$F_s = \frac{0.43 \times 177}{1.4} = 56.9 \text{ kN,}$$

the number required is

$$N = \frac{424}{56.9} = 7.45, \text{ say 8.}$$

The net area of each cover plate (allowing for two holes) is

$$A_n = T[300 - (2 \times 24)] = 252T$$

Using a tensile stress of $p_t = 155$ N/mm^2,

$$A_n = \frac{424 \times 10^3}{155} = 2735 \text{ mm}^2$$

giving

$$T = \frac{2735}{252} = 10.86, \text{ say 12 mm.}$$

Note: If the moment was a wind moment the bolt load factor would be 1.2 and p_t increased by 25 per cent in accordance with the specifications.

Provide a division plate and machine column ends as before. A detail is given in Fig. 8.

EXAMPLE 7.

A load of 110 kN is to be carried on a bracket formed from two 12 mm thick plates, arranged 250 mm apart and bolted to the webs of a pair of $254 \times 76 \times 28.29$ channels, giving an eccentricity of 300 mm. Design the bolt group.

SOLUTION

The eccentricity moment in this problem is in the plane of the bolts, and is approached as follows, subject to appreciating that the exercise is one of trial and error. The arrangement tested is shown in Fig. 9.

Using the simple torsion equation, $M_t/I_p = q/r$,

$$M_t = We = 55 \times 300 = 16\,500 \text{ kNmm}$$

$$I_p = A \sum (x^2 + y^2),$$

where A = area of each bolt, then

$$\sum y^2 = 4 \times 90^2 = 32\,400$$
$$+ 4 \times 30^2 = 3600$$
and
$$\sum x^2 = 8 \times 70^2 = 39\,200$$
$$\therefore \quad \sum (x^2 + y^2) = 75\,200$$

whence $I_p = 75\,200\,A$ mm^4

r = distance from C.G. of bolt group to bolt nearest load
$$= \sqrt{(90^2 + 70^2)} = 100\sqrt{1.3}.$$

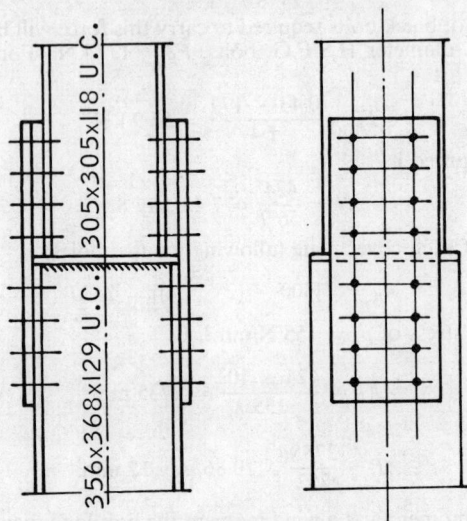

Flange covers	300 × 12 mm plate
Division plate	25 mm thick
Flange packs	thickness to suit
Bolts	22 mm dia. H.S.F.G.

Fig. 8

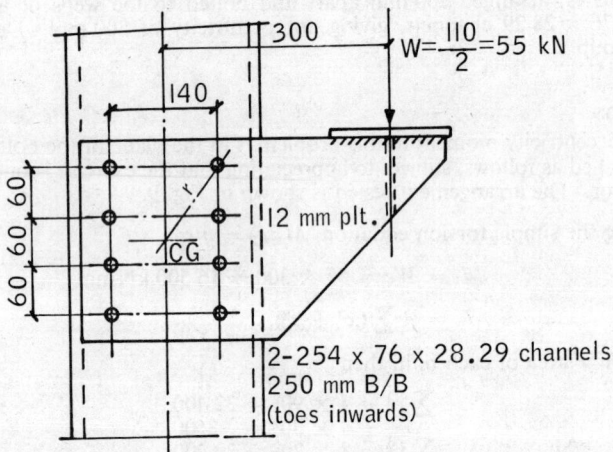

Fig. 9

Substitute into torsion equation to give

$$\frac{16\,500}{75\,200A} = \frac{q}{100\sqrt{1\cdot3}},$$

whence
$$Q = qA = 25 \text{ kN},$$
which is the turning force on the extreme bolt and is normal to the radial distance r.

The direct vertical force on each bolt is

$$V = \frac{W}{N} = \frac{55}{8} = 6.88 \text{ kN}$$

Resolving Q and V gives the maximum force on this bolt:

$$R = \sqrt{(Q^2 + V^2 + 2QV \cos \varphi)} \quad \text{where } \cos \varphi = x/r$$
$$= \sqrt{\{25^2 + 6.88^2 + [2 \times 25 \times 6.88 \times (70/100\sqrt{1.3})]\}} = 29.7 \text{ kN}$$

From the section tables the channel web is 8·1 mm thick. Using machined bolts 20 mm diameter shear governs giving a safe capacity of 31·4 kN; black bolts 24 mm diameter shear governs giving a safe capacity of 36·2 kN; friction-grip bolts 16 mm diameter have a safe capacity of 29·6 kN.

Checking the bracket plates,

$$Z = \frac{t_w d^2}{6} = \frac{12 \times 240^2}{6} = 115\,200 \text{ mm}^3,$$

$$A = t_w d = 12 \times 240 = 2880 \text{ mm}^2,$$

giving

$$f_{bc/t} = \frac{16\,500 \times 10^3}{115\,200} = 143.4 \text{ N/mm}^2,$$

$$f_q = \frac{55 \times 10^3}{2880} = 19.1 \text{ N/mm}^2$$

which are satisfactory.

EXAMPLE 8.

A load of 110 kN is to be supported on a bracket formed from a 356 × 127 × 39 U.B. which is bolted through an end plate to the flange of a 254 × 254 × 89 U.C.. The line of the load is 300 mm eccentric from the centre line of the column.

SOLUTION

The vertical load produces shear in the bolts and, because the line of the load is eccentric to the bolt group, tension will be induced. Assume an arrangement consisting of two rows of four bolts pitched vertically at 90 mm centres. Also assume the bolts are 20 mm diameter and the end plate is 150 mm wide.

The technique is to replace the bolt group and the end plate by an equivalent inverted T-section which may be operated upon by direct application of simple bending theory. Since the bolt spacing is regular, the width t is the area of two bolts divided by their vertical pitch. Stress area (tensile) of M20 bolt = 245 mm²

$$t = \frac{2 \times 245}{90} = 5.44 \text{ mm}$$

taking moments about the neutral axis, located at x from top

$$\frac{5.44x^2}{2} = \frac{150}{2}(360 - x)^2$$

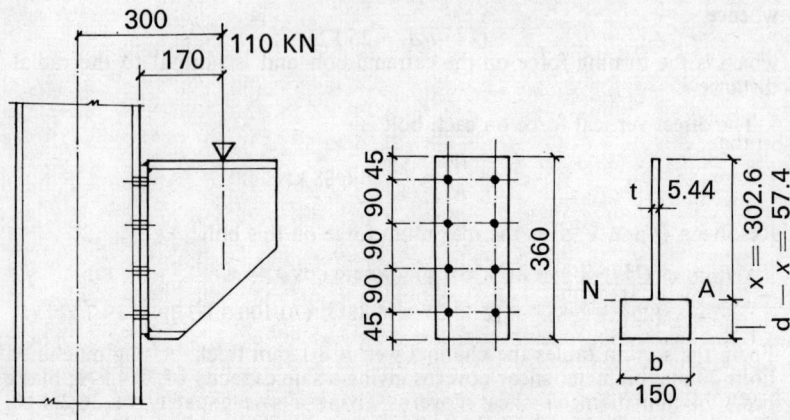

FIG. 10

which reduces to $\qquad 0\cdot964\,x^2 - 720x + 129\,600 = 0$

giving $\quad x = \dfrac{720 - \sqrt{720^2 - 4 \times 0\cdot964 \times 129\,600}}{2 \times 0\cdot964} = 302\cdot6$ mm

This indicates that the bottom two bolts are below the neutral axis *i.e.* in the compression zone. They do not contribute to the tension but will resist part of the shear.

The second moment of area I of the equivalent shape is

$$I_{\text{NA}} = \frac{5\cdot44 \times 302\cdot6^3}{3} + \frac{150 \times 57\cdot4^3}{3} = 59\,699\,985 \text{ mm}^4$$

and the distance from the neutral axis to the top bolts is

$$y = 302\cdot6 - 45 = 257\cdot6 \text{ mm}$$

Hence $\qquad Z_{\text{TOP}} = \dfrac{59\,699\,985}{257\cdot6} = 231\,755 \text{ mm}^3$

Distance from force to bolt line is $300 - \dfrac{260}{2} = 170$ mm

$$\therefore \text{ moment } M = 110 \times 170 \times 10^3 = 18\cdot7 \times 10^6 \text{ N mm}$$

At the level of the top bolts the 'bending' stress is

$$f = \frac{18\cdot7 \times 10^6}{231\,755} = 80\cdot7 \text{ N/mm}^2$$

this is also the direct tensile stress in the top bolts.
Permissible stress in tension on grade 4·6 bolts is 120 N/mm²

so that $\qquad \dfrac{f_t}{p_t} = \dfrac{80\cdot7}{120} = 0\cdot67 < 1$

Since eight bolts will resist shear, the shear stress will be (using the gross diameter of 20 mm)

$$f_s = \frac{110 \times 10^3}{8} \times \frac{4}{\pi \times 20^2} = 43.8 \text{ N/mm}^2$$

Permissible shear stress for Grade 4·6 bolts is 80 N/mm²

So that

$$\frac{f_s}{p_s} = \frac{43.8}{80} = 0.55 < 1$$

and

$$\frac{f_t}{p_t} + \frac{f_s}{p_s} = 0.67 + 0.55 = 1.22 < 1.4$$

An alternative method which is quicker to operate is shown below. It makes two assumptions:

1. the neutral axis lies through the lowest bolts;
2. all the applied moment is resisted by the bolts.

From the previous example, the modulus of the bolt group will be

$$Z = \frac{2A\Sigma Y^2}{y}, \text{ this gives}$$

$$Z = \frac{2A}{270} [270^2 + 180^2 + 90^2] = 840A \text{ mm}^3$$

If A = tensile stress area, 245 mm², the direct bolt stress is

$$f_t = \frac{18.7 \times 10^6}{840 \times 245} = 90.9 \text{ N/mm}^2$$

Using the same permissible stresses as before and noting that the shear stress of 43·8 N/mm² does not change, then

$$\frac{ft}{p_t} + \frac{f_s}{p_s} = \frac{90.9}{120} + \frac{43.8}{80}$$

$$= 1.31 < 1.4$$

EXAMPLE 9.

The forces acting at a beam to column connection are as follows:

> Reaction from vertical loads = 180 kN
> Reaction from wind loads = ±47·6 kN
> Moment from wind loads = ±190·4 kN m

Design a suitable connection using bolts. Both the beam and the column are grade 43 steel; the beam is a 533 × 210 × 92 U.B. and the column 305 × 305 × 118 U.C. For this connection the beam can be fixed to the column, using a pair of web cleats for shear and structural tees cut from U.B.s attached to the beam flanges to resist bending. The arrangement is shown in Fig. 11. All stresses may be increased by 25 per cent on account of wind in accordance with both clause 13 of B.S. 449 and B.S. 4604.

SOLUTION

Bolts between beam flange and stalk of tee—position A
    Force to be resisted is

$$F = \frac{M}{D} = \frac{190.4 \times 10^3}{533} = 357 \text{ kN}$$

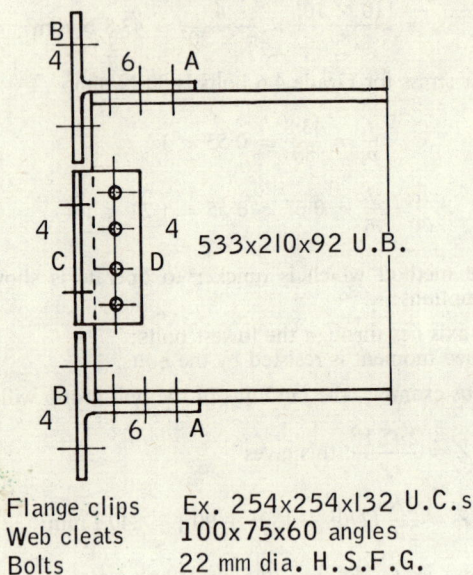

Flange clips   Ex. 254x254x132 U.C.s
Web cleats   100x75x60 angles
Bolts   22 mm dia. H.S.F.G.

FIG. 11

using 22 mm diameter friction-grip bolts $F_p = 177$ kN and a load factor of 1·2 (B.S. 4604),

$$F_s = \frac{0.45 \times 177}{1.2} = 66.4 \text{ kN}$$

number required

$$N = \frac{357}{66.4} = 5.4, \text{ say } 6.$$

Bolts between column flange and flange of tee—position B

Force to be resisted is 357 kN as before. Maximum permitted force on bolts which are now subject to direct external tension is $0.6F_p = 106.2$ kN, and the number required is

$$N = \frac{357}{106.2} = 3.4, \text{ say } 4.$$

Strength of tees to resist wind forces

Force to be resisted from wind moment = 357 kN as before, therefore net area of stalk required is

$$A_n = \frac{357 \times 10^3}{155 \times 1.25} = 1845 \text{ mm}^2$$

Make stalk 220 mm wide at first bolt position, less two 24 mm diameter holes, then the least thickness required is

$$t = \frac{1845}{[220 - (2 \times 24)]} = 10.7 \text{ mm}$$

The flange of the tee will be subject to bending, see Fig. 12.

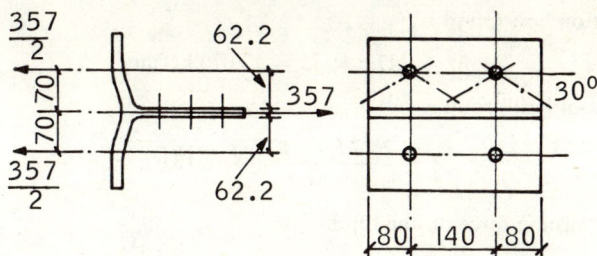

FIG. 12

For the bolting arrangement assume the spread lines overlap and allow the full width of the tee cutting to be used, *i.e.* 300 mm. Try a tee cut from a $254 \times 254 \times 132$ U.C., $T = 25 \cdot 1$ mm, $t = 15 \cdot 6$ mm,

$$Z \text{ of flange} = \frac{300 \times 25 \cdot 1^2}{6} = 31 \cdot 5 \times 10^3 \text{ mm}^3$$

Taking the flange as a fixed-ended beam then

$$M = \frac{357}{2} \times \frac{62 \cdot 2}{2} = 5551 \text{ kN mm}$$

$$\therefore f_{bc} = \frac{5551 \times 10^3}{31 \cdot 5 \times 10^3} = 176 \cdot 4 \text{ N/mm}^2$$

and

$$p_{bc} = 165 + 25\% = 206 \cdot 3 \text{ N/mm}^2$$

The tee cutting assumed is satisfactory since both flange and stalk thicknesses are adequate for the forces acting.

Web cleats

Without wind shear = 180 kN and with wind shear = 227·6 kN, since the latter exceeds the former by 26·5 per cent the 25 per cent stress increase is justified.

Bolts required between face of column and cleats—position C

Assuming 22 mm diameter friction-grip bolts $F_s = 66 \cdot 4$ kN as previously calculated, then number required is

$$N = \frac{227 \cdot 6}{66 \cdot 4} = 3 \cdot 43, \text{ say } 4.$$

Bolts required in web of beam

Between the face of the column and the end of the beam there is a gap of 40 mm due to the flange moment connections. Allowing for 35 mm edge distance the web bolts will be 75 mm eccentric, thus producing a moment from the reaction. Try four 22 mm diameter friction-grip bolts vertically spaced at 55 mm centres. Direct force per bolt,

$$F_d = \frac{227 \cdot 6}{4} = 56 \cdot 7 \text{ kN}$$

Moment on bolt group is

$$M = 227 \cdot 6 \times 75 = 17\,070\,\text{kN mm}$$

Modulus of bolt group is

$$Z = \frac{2(27 \cdot 5^2 + 82 \cdot 5^2)}{82 \cdot 5} = 183 \cdot 7,$$

giving a turning force on the top bolt of

$$F_t = \frac{17\,070}{183 \cdot 7} = 92 \cdot 8\,\text{kN}$$

Hence

$$F_R = \sqrt{(92 \cdot 8^2 + 56 \cdot 7^2)} = 108 \cdot 8\,\text{kN}$$

The "double-shear" value of the assumed bolt is $2 \times 66 \cdot 4 = 132 \cdot 8$ kN, therefore arrangement is adequate.

WELDED CONNECTIONS

One structural member may be connected to another by welding. However, this method should not be used indiscriminately otherwise conditions will obtain not intended by the designer. Generally, welding is performed by the electric arc process using consumable electrodes, and may be either manual or automatic. Electrodes and the metal being welded should be chemically compatible.

Welded joints may be made by fillet welds or butt welds. Parts being fillet welded need no edge preparation other than that necessary to ensure proper alignment. Parts being butt welded require edge preparation for the proper depositing of weld metal in addition to being in correct alignment. Three welded joints are shown in Fig. 13.

Welds and welding are covered by Part G of B.S. 449, and welding terms and symbols by B.S. 499. Other welding specifications of importance are B.S. 639 and B.S. 5135.

Permissible stresses in welds

Providing the electrodes comply with B.S. 639 the permissible stresses in fillet welds tabulated below may be used.

Steel grade B.S. 4360	Relative section in B.S. 639	Permissible stress, p_w
43	1 and 2	115
50	1 and 4	160
55	1 and 4	195

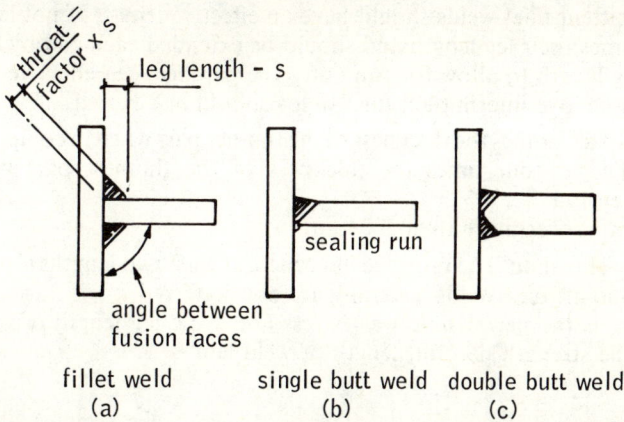

Fig. 13

The strength of fillet welds is related to the thickness (*see* Fig. 13) which in turn depends upon the angle between the fusion faces. This relationship is given in the following table.

Angle between fusion faces	60–90°	91–100°	101–106°	107–113°	114–120°
Factor by which leg length is to be multiplied to obtain throat thickness	0·70	0·65	0·60	0·55	0·50

The strength of butt welds is also related to the throat thickness (*see* Fig. 13) and may be treated as parent metal, but they must be sealed to obtain full strength. Unsealed single butt welds should have a throat thickness of at least seven-eighths of that shown, but for strength calculations only five-eighths may be taken.

Design of welds

Those three rules given for bolted connections apply equally for welded connections.

1. The product of the length of weld and its strength per unit run must be not less than the force it is required to resist.
2. The moment of resistance of a weld group must be not less than the moment it is required to resist.
3. The plate and/or sections employed to form the connection of which the weld forms part must be capable of safely resisting the forces and moments to which they are subject.

Intermittent fillet welds should have an effective length of not less than four times their leg length and should be extended each end by at least one leg length to allow for run out. The gap between effective lengths of consecutive intermittent fillet welds should not exceed:

1. Sixteen times the thickness of the thinner part when in compression;
2. Twenty-four times the thickness of the thinnest part when in tension;
3. but never more than 300 mm.

Fillet welds should be returned at corners at least two lengths if possible. Intermittent butt welds generally are not preferred.

If p_w is the permissible weld stress and s the leg length of the weld then the strength per unit length of weld laid is

$$F_s = p_w k s,$$

where k is the fusion face angle factor from the table on p. 21.

EXAMPLE 10

The reactions from a $457 \times 152 \times 82$ U.B. in grade 43 steel are 100 kN and 300 kN respectively. Design the bottom brackets which are to be welded to the flanges of the receiving columns.

SOLUTION

100 kN reaction

The bearing capacity of the beam web, without stiff bearing, is capable of resisting this reaction; therefore only a simple bracket is required. This may be cut from a $125 \times 75 \times 10$ angle long leg welded to column flange. Assuming two fillet welds of 6 mm leg length, and a permissible stress $p_w = 115$ N/mm², length of weld required is

$$l = \frac{100 \times 10^3}{2 \times 115 \times 0 \cdot 7 \times 6} = 103 \cdot 5 \text{ mm}$$

Alternatively, assuming 10 mm run-in at cleat top, but returned along bottom, effective weld length is

$$l = 125 - 10 = 115 \text{ mm},$$

the leg length required is

$$s = \frac{100 \times 10^3}{2 \times 115 \times 117 \times 0 \cdot 7} = 5 \cdot 3, \text{ say 6 mm.}$$

The proposed connection is shown in Fig. 14.

300 kN reaction

To achieve the bearing capacity the beam requires 100 mm of stiff bearing, which means that the support bracket must be stiffened and have an outstand of at least 100 mm. The reaction will now be eccentric to the face of the column by at least 50 mm. Try a short length of $229 \times 305 \times 51$ tee—which will be attached to the column as shown in Fig. 15.

The bending moment on the weld group is

$$M = 300 \times 50 = 15\ 000 \text{ kN mm}$$

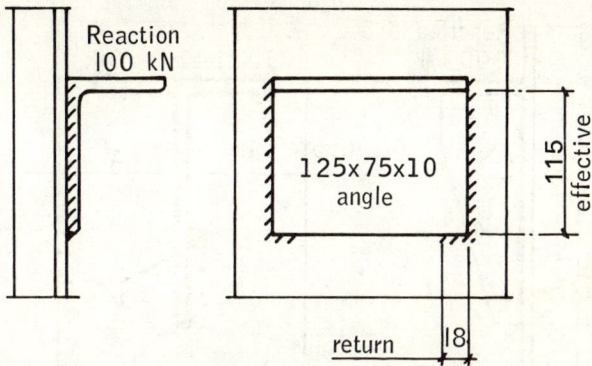

FIG. 14

Assuming the weld has unit leg thickness find the C.G. about the top weld, then

Part	Weld length	Lever arm	
Top flange	30×1	0	0
Under flange	217×1	14·8	3211·6
Flange toes	$14·8 \times 2$	7·4	219·0
Web	$286·3 \times 2$	157·9	90 413·5
Sum	849·2		93 844·1

therefore

$$\bar{y} = \frac{93\,844·1}{849·2} = 110·5 \text{ mm.}$$

About the C.G. the moment of inertia will be

Part	Weld length	y	$l \times y^2$
Top flange	30×1	110·5	366 450
Under flange	217×1	95·7	1 987 394
Flange toes	$14·8 \times 2$	103·1	314 648
Web	$286·3 \times 2$	47·4	1 286 518
Web	$\dfrac{2 \times 286·3^2}{12}$		3 909 826
Sum			7 864 836

therefore

$$\text{weld modulus } Z_w = \frac{7\,864\,836}{110·5} = 71\,175$$

The "stresses" due to the bending moment and the direct force will be

$$f'_{bt} = \frac{15\,000 \times 10^3}{71\,175} = 210·7 \text{ N/mm,}$$

$$f'_d = \frac{300 \times 10^3}{849·2} = 353·2 \text{ N/mm,}$$

giving a resultant "stress" of

$$f'_R = \sqrt{(210·7^2 + 353·2^2)} = 411·3 \text{ N/mm,}$$

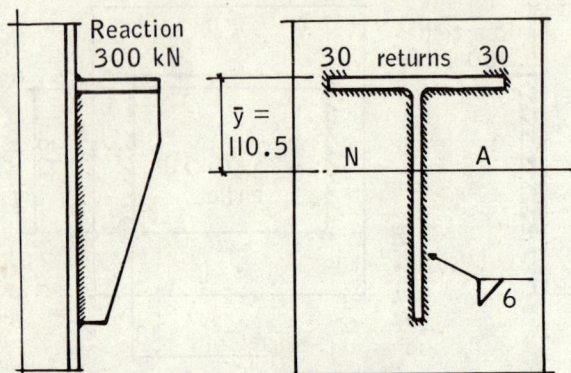

Stiff bracket   Ex. 229x305x51 tee x 100 long
welded as shown

FIG. 15

at a weld stress $p_w = 115$ N/mm²,

$$411 \cdot 3 = 115 \times 0 \cdot 7s$$
$$\therefore \; s = \frac{411 \cdot 3}{115 \times 0 \cdot 7} = 5 \cdot 1, \text{ say } 6 \text{ mm.}$$

EXAMPLE 11

A load of 250 kN is to be carried on a welded bracket formed from two 12 mm plates connected to the webs of a pair of $254 \times 76 \times 28 \cdot 29$ channels arranged toe to toe. The load is eccentric to the centre line of the channels by 300 mm. Design the welding.

SOLUTION

The eccentricity moment in this problem is in the plane of the weld group and is similar to the bolt group problem given in Example 7. Try the arrangement in Fig. 16.

Using the simple torsion equation $M_t/I_p = q/r$, the eccentricity of the load measured from the centroid of the weld group is 293 mm, hence the moment in each plate is

$$M_t = We = \frac{250 \times 293}{2} = 36\,625 \text{ kN mm}$$

Now the polar moment of inertia of the weld group is $I_p = I_x + I_y$, and since the group is symmetrical, then for unit thickness,

$$
\begin{aligned}
I_x = & \;\; 2 \times 240 \times 150^2 &= 10\,800\,000 \\
& +2 \times 300^3/12 &= 4\,500\,000 \\
I_y = & \;\; 2 \times 300 \times 120^2 &= 8\,640\,000 \\
& +2 \times 240^3/12 &= 2\,304\,000 \\
\hline
I_p = & \;\; I_x + I_y &= 26\,244\,000
\end{aligned}
$$

And the distance to the extreme corner

$$r = \sqrt{150^2 + 120^2} = 192 \cdot 1$$

giving the weld modulus as

$$Z_w = \frac{26\,244\,000}{192 \cdot 1} = 136\,621$$

Hence the "bending" stress is

$$f_b = \frac{36\,625 \times 10^3}{136\,621} = 268 \cdot 1 \ \text{N/mm}$$

The total effective weld length is

$$l = 2\,(300 + 240) = 1080 \ \text{mm}$$

giving a "direct" stress of

$$f_d = \frac{240 \times 10^3}{2 \times 1080} = 115 \cdot 7 \ \text{N/mm}$$

so that the resultant stress is

$$f_r = \sqrt{115 \cdot 7^2 + 268 \cdot 1^2 + (2 \times 115 \cdot 7 \times 268 \cdot 1 \times 120/192 \cdot 1)}$$
$$= 352 \cdot 2 \ \text{N/mm}$$

using a weld stress of $p_w = 115 \ \text{N/mm}^2$, then

$$352 \cdot 2 = 115 \times 0 \cdot 7 \times s$$

from which $\qquad s = \dfrac{352 \cdot 2}{115 \times 0 \cdot 7} = 4 \cdot 37 \ \text{mm},$

say 5 mm fillet weld.

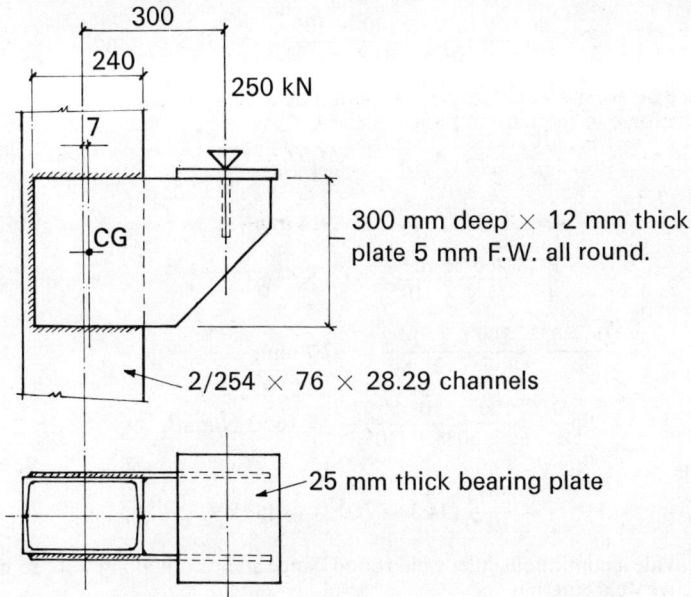

FIG. 16

Note: From more advanced torsion theory the modulus of the weld group may be found from the expression

$$Z_w = 2At$$

where A equals the area enclosed by the boundary of the weld metal and t is the weld leg thickness.

Assuming unit weld thickness $t = 1$, then

$$Z_w = 2 \times 300 \times 240 = 144\,000 \text{ mm}^3$$

from which the torsional stress is

$$f_s = \frac{36\,625 \times 10^3}{144\,000} = 254 \cdot 3 \text{ N/mm}$$

and the resultant would be 338·8 N/mm which would produce a weld leg thickness of 4·21 mm.

EXAMPLE 12

A 203 × 203 × 60 U.C. is attached symmetrically to the end of a 914 × 305 × 201 U.B. to form a valley line double-crane column in a two-bay workshop. The U.C. resists an axial force of 260 kN, plus a major axis moment of 100 kN m due to an eaves shear force of 25 kN. There is an additional shear at the junction of the two members due to crane action of 60 kN. Design the connection assuming grade 50 materials.

SOLUTION

Welding between U.C. and division plate

Check for tension in the U.C.

$$f = \frac{260 \times 10^3}{75 \cdot 8 \times 10^2} - \frac{100 \times 10^6}{581 \cdot 1 \times 10^3} = -137 \cdot 9 \text{ N/mm}^2$$

Hence determine weld size by calculation as tension develops.

The force in the tension flange of the U.C. is

$$F_t = A_f\left[\frac{P}{A} - \frac{M}{I}\left(\frac{D-T}{2}\right)\right],$$

where

$$A_f = 205 \cdot 2 \times 14 \cdot 2 = 2914 \text{ mm}^2,$$

$$\frac{P}{A} = \frac{260 \times 10^3}{75 \cdot 8 \times 10^2} = 34 \cdot 3 \text{ N/mm}^2,$$

$$\frac{D-T}{2} = \frac{209 \cdot 6 - 14 \cdot 2}{2} = 97 \cdot 7 \text{ mm},$$

$$\frac{M}{Z'} = \frac{100 \times 10^6 \times 97 \cdot 7}{6088 \times 10^4} = 160 \cdot 3 \text{ N/mm}^2,$$

giving

$$F_t = \frac{2914}{1000}(34 \cdot 3 - 160 \cdot 3) = 368 \text{ kN}$$

Provide a continuous fillet weld round flange and return along web, 25 mm. Effective weld length is

$$l = 205 \cdot 2 + (205 \cdot 2 - 9 \cdot 3) + 2(14 \cdot 2 + 25) = 479 \cdot 5 \text{ mm}$$

Hence the force per unit run of weld is

$$F_w = \frac{368 \times 10^3}{479 \cdot 5} = 768 \text{ N/mm}$$

Using a weld stress of $P_w = 160 \text{ N/mm}^2$, then

$$768 = 160 \times 0 \cdot 7s$$

$$\therefore \ s = \frac{768}{160 \times 0 \cdot 7} = 6 \cdot 9, \text{ say 8 mm.}$$

Take the shear force of 25 kN on a pair of 50 mm long fillet welds laid along the web, then

$$25 \times 10^3 = 160 \times 2 \times 50 \times 0 \cdot 7s$$

$$\therefore \ s = \frac{25 \times 10^3}{160 \times 2 \times 50 \times 0 \cdot 7} = 2 \cdot 2, \text{ say 5 mm}$$

Welding between division plate and U.B.

Owing to the great difference in depth between the U.C. and U.B., assume that the forces and moments from the U.C. are accommodated locally. Four re-entrant stiffeners should be provided, one beneath each U.C. flange outstand to take the bending out of the division plate. Make these stiffeners 10 mm thick with an outstand of 100 mm, 80 mm of which are effective after deducting 20 mm for weld clearance, etc. Allowing for 30° spread take the web welds as 370 mm long.

The inertia of this welding arrangement is:

$I_x = 2$ web welds	$\dfrac{2 \times 370^3}{12}$	$=$	8442×10^3
2 outer flange welds	$4 \times 80 \times 105^2$	$=$	3528×10^3
2 under flange welds	$4 \times 80 \times 95^2$	$=$	2888×10^3
			$14\,858 \times 10^3$

$$\therefore \ Z_w = \frac{14\,858 \times 10^3}{185} = 80 \cdot 3 \times 10^3,$$

giving a bending "stress" of

$$f_b' = \frac{100 \times 10^6}{80 \cdot 3 \times 10^3} = 1246 \text{ N/mm}$$

The total weld length is

$$l = 2 \times 370 + 8 \times 80 = 1380 \text{ mm},$$

giving a direct vertical "stress" of

$$f_v' = \frac{260 \times 10^3}{1380} = 188 \cdot 5 \text{ N/mm}$$

Hence the net stress on the tension side of the weld group is

$$f_t' = 1246 - 188 \cdot 5 = 1057 \cdot 5 \text{ N/mm}$$

Taking the shear force on the web welds only,

$$f'_q = \frac{25 \times 10^3}{2 \times 370} = 33 \cdot 8 \text{ N/mm},$$

$$\therefore f'_R = \sqrt{(1057 \cdot 5^2 + 33 \cdot 8^2)} = 1064 \text{ N/mm}$$

Using a weld stress of $p_w = 160$ N/mm^2,

$$1064 = 160 \times 0 \cdot 7s,$$

$$\therefore s = \frac{1064}{160 \times 0 \cdot 7} = 9 \cdot 5, \text{ say 10 mm.}$$

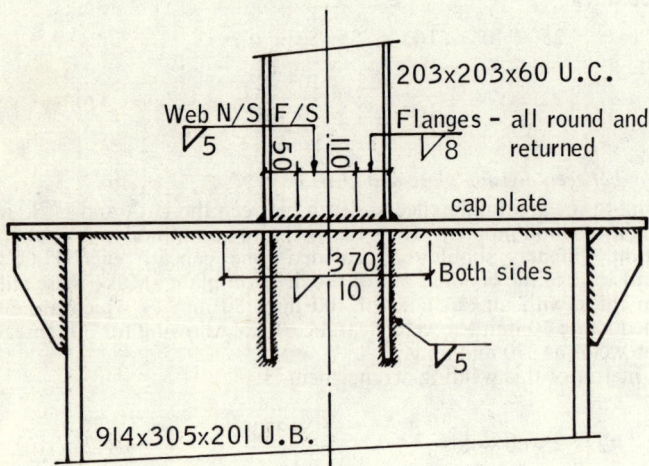

FIG. 17

Make the re-entrant stiffeners extend 200 mm down the web of the U.B. and provide weld to resist almost 368 kN on one pair of stiffeners each having two welds, then

$$368 \times 10^3 = 160 \times 2 \times 2 \times 0 \cdot 7s \times (200 - 20)$$

giving

$$s = 4 \cdot 6, \text{ say 5 mm.}$$

The shear of 60 kN from crane action can be easily taken up on the remaining U.B. profile. A nominal weld will suffice. Details of this column junction are given in Fig 17.

MIXED CONNECTIONS

Weld metal should not be used to increase strength in bolted connections, the opposite also applies. The reason is that in a black bolted connection the load is taken up after slip and the bolts are in contact with the surrounding metal: welding would prevent this action. In fact the weld metal would need to fail or seriously elongate to allow the bolts to operate and if deficient in number they may themselves fail.

Both machined bolts in close-fitting holes and friction-grip bolts do not take up their load by slipping, but their mechanical behaviour is doubtful if weld metal is used as an additive to strength.

Permissible stresses and design requirements are the same as those previously discussed.

EXAMPLE 13

Redesign the beam to column connection in Example 9 (p. 17) using an end plate welded to the beam and bolted to the column. The arrangement of this detail is shown in Fig. 18.

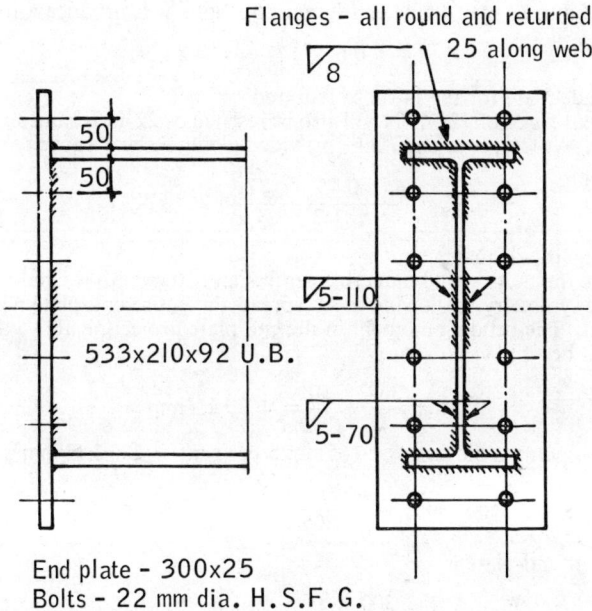

Flanges - all round and returned
25 along web

End plate - 300x25
Bolts - 22 mm dia. H.S.F.G.

FIG. 18

SOLUTION

End plate to beam

Using a simplified approach which avoids calculating the properties of the weld as in previous examples, the flange force to be resisted by the welds is

$$F = \frac{M}{D - T} = \frac{190\cdot4 \times 10^3}{533\cdot1 - 15\cdot6} = 368 \text{ kN}$$

Welding all round the flange and returning down web 25 mm gives a weld length of

$$l = 209\cdot3 + (209\cdot3 - 10\cdot2) + 2(15\cdot6 + 25)$$
$$= 489\cdot6 \text{ mm}$$

If stress is increased by 25 per cent to allow for wind, then

$$368 \times 10^3 = 1\cdot25 \times 115 \times 489\cdot6 \times 0\cdot7s,$$
$$\therefore \ s = 7\cdot5, \text{ say 8 mm.}$$

The shear force of 227·6 kN is taken by web welds. Assuming 5 mm fillet welds, one each side of web, then the length required is

$$l = \frac{227·6 \times 10^3}{1·25 \times 115 \times 0·7 \times 5 \times 2} = 226 \text{ mm}$$

Make this up of two lengths of 63 mm running on from the return of the flange weld, and one length of 100 mm symmetrical about the xx-axis. This gives a gross weld gap of 113 mm and complies with clause 54 of B.S. 449.

End plate to column

The force to be resisted by the bolts required to balance the moment will be 368 kN. Four 22 mm diameter friction-grip bolts will produce a force of

$$F = 4 \times 0·6 \times 177 = 424·8 \text{ kN}$$

which is adequate for the purpose required.

The shear force of 227·6 kN will also be resisted by 22 mm diameter friction-grip bolts. At least four should be provided giving a shear capacity of

$$F_s = \frac{4 \times 0·45 \times 177}{1·2} = 265 \text{ kN}$$

which is again adequate.

Arrange the bolts at 140 mm cross centres and 50 mm above and below the flange for symmetry. Allowing a 30° spread the entire end-plate width may be utilised. The bending moment in the end plate projecting above the beam flange will be

$$M = \frac{368}{2} \times \frac{50}{2} = 4600 \text{ kNmm.}$$

Taking the permissible stress at $165 + 25$ per cent $= 206·3$ N/mm²,

$$\text{required } Z = \frac{4600 \times 10^3}{206·3} = 22\,300 \text{ mm}^3$$

Modulus of end-plate is

$$Z = \frac{300 \times t^2}{6} = 50t^2,$$

then

$$50t^2 = 22\,300,$$

giving

$$t = \sqrt{\left(\frac{22\,300}{50}\right)} = 21·1, \text{ say 25 mm.}$$

CHAPTER 2

BEAMS AND GIRDERS

A structural member is termed a "beam" or "girder" when the loading it carries is resisted by bending action. From the elementary theory of bending is deduced the expression

$$\frac{M}{I} = \frac{f}{y} = \frac{E}{R}$$

which presupposes that the beam is bent into a circular deflected shape due to a uniformly applied bending moment. In practice this is rarely the case, since the various loading conditions met with in everyday design always produce a bending-moment pattern which varies in some manner along the length of the beam. However, since the ratio of span to depth is usually large the above expression may be regarded as quite reliable. The total design of a beam or girder involves checking stress levels from various effects and ensuring that the deflection is within some prescribed limit. The following stress conditions should be investigated:

(a) bending,
(b) web shear,
(c) web bearing,
(d) web buckling,
(e) combined bending and web shear,
(f) combined bending, web shear and web bearing.

None of these stress levels should exceed the allowable values given in the design specification. Although beams and girders perform a similar function, their make-up is different. The former are profiles produced at the rolling mills, for example universal beams and joists, although channels and indeed even angles and tees may be used as beams. The latter, as the name may imply, suggests a section of much larger proportions, which is obtained by welding together three or more plates to produce a beam-type profile.

Because the design approach for a beam and a plate girder is not entirely similar each will be treated separately.

DESIGN OF ROLLED BEAMS

Bending stresses

Having determined the loading the first step in the design process is to calculate the reactions, shears and maximum bending moments to be resisted. Since the actual size of the beam in the majority of cases will be determined by bending, the allowable compressive bending stress must be established. Unless the beam profile is unsymmetrical the tensile bending stress will take the same value, hence the compressive stress will be the criterion. This stress will depend upon the degree of restraint given to the compression flange, and the method of providing torsional restraint at the supports.

Clause 26 of B.S. 449 gives conditions of restraint for simply supported beams and cantilevers and for beams having cantilever ends. These conditions are expressed in terms of the effective length of the compression flange. No conditions are given for multi-bay beams designed to be fully continuous. When the entire length of the compression flange is built into a floor slab or is attached at frequent intervals by bolting or welding to floor plates it is considered fully restrained. Alternatively, where a floor slab is capable of providing frictional resistances from its self-mass equal to at least $2\frac{1}{2}$ per cent of the flange force in the beam then full restraint may be assumed. If this frictional resistance is inadequate positive connectors should be employed. Failing this, the degree of restraint must be assessed in relation to the end conditions of the beam. Given that torsional end restraint is obtained from flange cleats, web cleats or bearing stiffeners or that the ends of the beam are solidly built into walls, effective lengths to be employed in determining allowable bending stresses may be taken from the following table.

Beam type	Terminal restraints	Effective length
Simple beam	Ends of compression flanges unrestrained against lateral bending	$l = $ span
	Ends of compression flanges partially restrained against lateral bending	$l = 0{\cdot}85 \times$ span
	Ends of compression flanges fully restrained against lateral bending	$l = 0{\cdot}7 \times$ span
Cantilever	Built in at support, but free at end	$l = 0{\cdot}85 \times$ span
	Built in at support, but restrained against torsion at end by trimmer beam	$l = 0{\cdot}75 \times$ span
	Built in at support, but restrained against torsion and lateral displacement at end by trimmer beams extending over several bays, some of which are braced	$l = 0{\cdot}50 \times$ span
Cantilevers (in beam and cantilever const.)	Continuous, but unrestrained torsionally at support and free at end	$l = 3 \times$ span
	Continuous, but partially torsionally restrained at support and free at end	$l = 2 \times$ span
	Continuous, but torsionally restrained at support and free at end	$l = $ span

Note: For beams, span means the effective span taken between assumed points of support; for cantilevers, span means the projection of the cantilever beyond the point of support.

When a floor system consisting of main and secondary beams carries a steel deck of the open-grill type the deck is usually located at infrequent intervals, and provides little or no restraint to the beams. The effective length of the main beams will depend upon the spacing of the secondaries, which themselves will in all likelihood be unrestrained. Two and a half per cent of the maximum flange force of the beam restrained should be shared equally by the elements providing restraint. Elements providing torsional restraints should resist the whole $2\frac{1}{2}$ per cent.

The theory from which the allowable compressive bending stress is deduced is outside the scope of this book. However, a glance at Table 3 of B.S. 449 shows that it is a function of two non-dimensional parameters termed l/r_y and D/T. The first of these quantities is the effective length of the compression flange divided by the radius of gyration about the yy-axis, and the second is the ratio of the beam depth to flange thickness. This means that some knowledge of the size of the beam is required beforehand which implies guesswork, but this is not so. Proceed first on the assumption that the compression flange is fully restrained, *i.e.* $l/r_y = 0$, using the maximum allowable bending stress to find the smallest section modulus, and select a beam from the section tables to suit. Clearly this beam is the smallest which may be considered. If the slenderness ratio l/r_y is less than any tabular value the permissible stress will be that previously assumed. If not then proceed as follows:

1. take the beam first selected or another somewhat stronger, never weaker for obvious reasons, and calculate the actual bending stress, namely;

$$f_{bc} = \frac{\text{actual bending moment}}{\text{selected section modulus}};$$

2. determine the slenderness ratio l/r_y and the D/T ratio and obtain p_{bc} from Table 3;
3. clearly $f_{bc} \not> p_{bc}$;
4. If $f_{bc} > p_{bc}$ move up the beam table until a section is found which just works.

Checking for bending is now complete.

Occasionally a situation will arise where the compression flange needs to be larger than the tension flange. Using rolled beams, this will call for a compounded compression flange which is obtained either by welding on a plate or a channel section. Because of the imposed lack of symmetry the stresses in both flanges must be checked. If I_c is the compounded second moment of area and y_c and y_t are the distances from the neutral axis then

$$f_{bc} = \frac{M}{I_c} y_c \not> p_{bc},$$

$$f_{bt} = \frac{M}{I_c} y_t \not> p_{bt}$$

The value of p_{bc} comes from Table 3 as previously described, or is the maximum value from Table 2 depending upon proportions. p_{bt} is not dependent upon l/r_y and takes the value given in Table 2.

Beams supporting overhead travelling cranes and angles employed as beams may be looked upon as special cases.

Deflection

Having found that the beam is satisfactory for bending stress, it is better if at this stage it is checked against deflection requirements. If the beam has a standard loading pattern the appropriate deflection equation should be used. Since the limiting deflection is specified there is no need to calculate the actual deflection, instead transpose the equation and find the least second moment of area required. The section selected for bending should also meet this requirement.

For simply supported beams resisting a system of loads which produce a bending moment tending towards a parabola, a first approximation of suitability against deflection may be obtained if the actual span to depth ratio L/D is less than that obtained from the expression

$$\frac{L}{D} = \frac{2800}{kf}$$

in which f is the actual bending stress due to the total load carried and k is the ratio of the load considered for deflection purposes to the total load carried.

The above expression is derived from clause 15 of B.S. 449 which requires that the maximum deflection should not exceed 1/360 of the span due to loads other than the weight of the structural floors or roof, steelwork and casing if any. Clearly, if the loading is uniformly distributed over the whole span the bending moment will be parabolic and the above expression is exact.

Shear stresses

The actual average shear stress is obtained by dividing the maximum shearing force by the cross-sectional area of the beam web. The stress obtained should not exceed the value obtained from Table 11 of B.S. 449, that is

$$f'_q = \frac{Q}{Dt_w} \ngtr p'_q$$

The actual maximum shear stress is obtained from the formula

$$f_q = \frac{QA\bar{Y}}{It_w} \ngtr p_q$$

which should not exceed the value obtained from Table 10. f_q will be a maximum at the neutral axis of the beam.

Shaded area = A

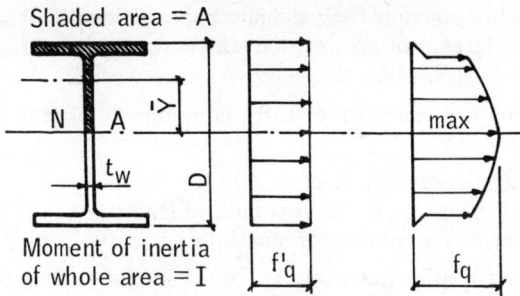

Moment of inertia
of whole area = I

FIG. 19

These two shear stress conditions are illustrated in Fig. 19.

Should either f_q' or f_q exceed the permissible value, two courses of action are available;

1. increase the size of the beam;
2. provide thickening plates.

Increasing the size of the beam is by far the simplest as this may only involve increasing the weight of the beam within a given serial size, i.e. 203 × 133 × 25 U.B. increased to 203 × 133 × 30 U.B. Use of thickening plates is not always economic since, subject to the nature of the applied loads, these plates may be excessively long compared to the beam span. They should therefore be employed only where the shear force is locally very high. On no account will the provision of web stiffeners improve the shear capacity of rolled beam sections. The reason behind this statement may be explained as follows. The permissible shear stress p_q is derived with regard to the level of critical shear stress which would produce a state of shear instability in the beam web. This critical shear stress is a function of two parameters, the ratio of web depth to thickness d/t and the ratio of unstiffened web length to depth a/d. Rolled beams are manufactured such that d/t is small (generally less than 60) which permits a/d to tend to a maximum, i.e. infinity, giving a high critical shear stress in excess of the shear yield stress of the material. Since the latter is thus the criterion, this is factored to give the permissible shear stress.

Bearing stresses

At positions of concentrated load local high compressive stresses are produced at the root of the beam web. These compressive stresses are termed "bearing stresses." Two positions occur: the first is at the support where the reaction produces the local bearing stress; and the second is under point loads within the span (loads from posts carried on the top flange, or incoming beams). The resulting bearing stresses must be limited to guard against local crushing of the beam web. It is assumed that this concentrated load or reaction will spread from its point of application at an angle of 30° to the horizontal up to the root of the beam web. At this position the calculated spread length of the load should produce a stress not greater than the permissible bearing stress given in Table 9 of B.S. 449.

The spread length is made up of three components of length which are:

1. the stiff bearing component of length, l_1;
2. the flange or bearing plate component of length, l_2;
3. the web bearing component of length, l_3.

These components are illustrated in Fig. 20 for simple bearing support.

For end bearing supports it is usual to leave erection clearance of about

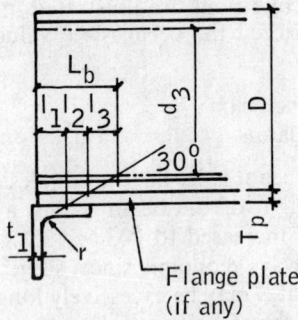

Fig. 20

5 mm between the end of the beam and the face of the column (web cleats if any will project beyond the end of the beam). Then the bearing length is

$$L_b = l_1 + l_2 + l_3,$$

where

$$l_1 = t_1[\sqrt{(3)} + 1] + r[\sqrt{(3)} - 1] - 5 \not> \frac{D}{2},$$

If the root radius r is put equal to the angle thickness, t_1, then the stiff bearing component will be approximately

$$l_1 = 3t_1$$

$$l_2 = T_p\sqrt{3},$$

$$l_3 = \frac{D - d_3}{2}\sqrt{3}$$

If the thickness of the beam web is t_w then the bearing area is

$$A_b = L_b t_w$$

For a given reaction, R

$$f_b = \frac{R}{A_b} \not> p_b$$

For intermediate or continuous supports over a column cap plate

$$L_b = l_1 + 2l_2 + 2l_3,$$

where l_1 = breadth of column cap plate $\geqslant D$, and l_2 and l_3 have the values previously assigned.

Note: Should the projection of the cap plate beyond the face of the column exceed $T_c\sqrt{3}$, then any greater amount should be ignored unless the cap is stiffened.

Knowing L_b, the bearing stress f_b is found as in the simple case. Normally universal beams are not plated so that $l_2 = 0$. However, if the bearing proves critical this may be improved by providing a horizontal pack or bearing plate between the bracket and the underside of the beam when T_p would then be the thickness of this plate. If bearing is still critical the component l_1 may be increased by stiffening the bracket.

Buckling stresses

Web buckling stresses are also produced at positions of concentrated load: two positions occur as described for web bearing. Again it is assumed that this concentrated load or reaction will spread from its point of application, but at an angle of 45° to the horizontal up to the neutral axis of the beam web. The spread length of the load should produce a stress not greater than that given in Table 17 of B.S. 449. The spread length is made up of three components of length similar to bearing and are illustrated in Fig. 21 for simple support.

For end supports taking the same erection clearance, the length to be considered for buckling is

$$L_{bu} = l_1 + l_2 + l_3,$$

where
$$l_1 = 2t_1 + r(2 - \sqrt{2}) - 5 \geqslant \frac{D}{2},$$

Making assumptions similar to that for bearing

$$l_1 = 2t_1$$
$$l_2 = T_p,$$
$$l_3 = \frac{D}{2}$$

If the thickness of the beam web is t_w as before then the buckling area is

$$A_{bu} = L_{bu}t_w$$

and for a given reaction, R

$$f_{bu} = \frac{R}{A_{bu}} \geqslant p_c$$

For intermediate or continuous supports the buckling length

$$L_{bu} = l_1 + 2l_2 + 2l_3,$$

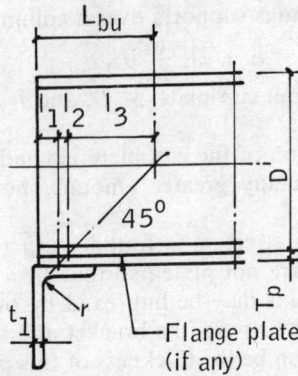

FIG. 21

where l_1 = breadth of column cap plate $\not> D$, and l_2 and l_3 have the values as assigned for the end support case. Since L_{bu} and t_w are known, the buckling stress f_{bu} may be found as above.

Now the permissible buckling stress for both end and intermediate supports depends upon the ratio of slenderness of the web. The procedure adopted is to treat the net depth of the web d_3 (this is the clear distance between the root fillets) as a column of effective height $0.5d_3$. Considering unit length of web in the direction of the beam span, then for a thickness of t_w, clearly

$$I_w = \frac{1 \times t^3_w}{12},$$

$$A_w = 1 \times t_w,$$

$$\therefore r = \frac{t_w}{2\sqrt{3}}$$

Hence the ratio of slenderness l/r is

$$\frac{l}{r} = \frac{0.5d_3}{r} = \frac{d_3}{t_w}\sqrt{3}$$

If the permissible stress obtained from Table 17 of B.S. 449 for this l/r is less than f_{bu}, then web stiffeners must be provided in accordance with Clause 28a (iii).

The simple rule given above holds good providing the flange through which the load or reaction is applied is effectively restrained against lateral movement relative to the other flange, and that the rotation of the loaded flange relative to the web is prevented. In default of these provisions the slenderness ratio shall be suitably increased.

Combined bending and web shear

At the supports of continuous beams and simple beams with cantilever ends there exists a combined effect of both bending stress and shear stress. This combined effect is termed equivalent stress. Beams having heavy concentrated point loads within their span will produce a similar situation, but this is not usually a criterion. From the elementary theory of bending, it is evident that at the point of maximum bending stress the shear stress is zero, and where the shear stress is a maximum the bending stress is zero. These two points correspond to the extreme fibre and the neutral axis respectively. In between both stresses will not be zero. Checking the combined stress at a point corresponding to the plane considered for bearing should produce the most adverse effect, since above this plane the bending stress is diminishing at a greater rate than the shear stress is increasing. This is shown at Fig. 22.

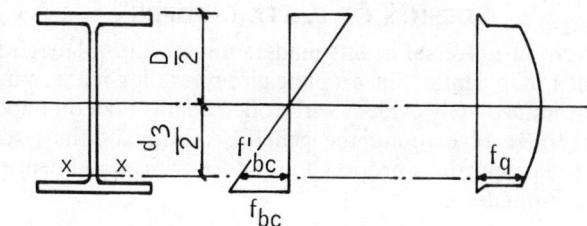

FIG. 22

With the case of a symmetrical beam at the plane considered, *i.e.* xx, this bending stress is

$$f'_{bc} = f_{bc} \times \frac{d_3}{D},$$

and the shear

$$f_q = \frac{QA\overline{Y}}{It_w}$$

Where A is the area between the plane xx and the extreme fibre and $\overline{Y}$ is the distance from the neutral axis to the centroid of this area.

These two stresses should be combined to give

$$f_e = \sqrt{[(f'_{bc})^2 + 3(f_q)^2]} \not> p_e$$

in which p_e is the permissible equivalent stress from Table 1 of B.S. 449. However, it should be noted that since $f'_{bc} < f_{bc}$ and f_q is less than the maximum as discussed under the heading "shear stresses" it is unlikely that f_e will exceed p_e.

Note: Where appropriate the tensile bending stress should be used.

Combined bending, shear and web bearing

Finally, the combined effect of stresses due to bending, shear and bearing should be investigated. These will obtain at supports mentioned in the previous section. The plane to consider is as for bending and shear. Since the reduced bending stress f'_{bc} has already been discussed, as has the shear stress f_q and the bearing stress f_b, all that is necessary is to combine these stresses to obtain the equivalent stress, *i.e.*

$$f_e = \sqrt{[(f'_{bc})^2 + (f_b)^2 - f'_{bc}f_b + 3(f_q)^2]},$$

if the bending stress is tensile then

$$f_e = \sqrt{[(f'_{bt})^2 + (f_b)^2 + f_{bt}f_b + 3(f_q)^2]} \not> p_e.$$

It will be clear that if all stresses are approaching their maximum values the equivalent stress f_e is likely to prove critical.

DESIGN OF PLATE GIRDERS

The advent of universal beams renders unnecessary plate girders less than about 1 m in depth. Since a plate girder is tailor-made, many more factors not immediately evident with rolled beams take on importance. These factors revolve round the general stability of the girder, and therefore the proportions proposed are more critical and hence subject to closer examination.

Basic proportions

Span-to-depth ratios are usually of a lower order than for rolled beams: a ratio of between 10 and 20 is reasonable for simply supported girders and up to about 35 for continuous girders. A breadth of one-quarter to one-third the depth leads to an ideal cross-section. The total area of the two flanges should be about twice the web area.

Detailed proportions

Local instability of flange plates must be avoided. For this reason flange breadth-to-thickness ratios are restricted. This ratio depends upon whether the flange is compressive or tensile, the grade of steel envisaged and edge stiffening. The table below gives the projection of the flange plate beyond the face of the web to satisfy local stability requirements.

Steel grade B.S. 4360	Max. projection of flange in terms of thickness, T			
	Unstiffened edges		Stiffened edges	
	Compression	Tension	Compression	Tension
43	16T	20T.	20T	20T
50	14T	20T	20T	20T
55	12T	20T	20T	20T

Note: This table relates to Clause 27b of B.S. 449.

Based upon similar considerations, the minimum thickness of the web plate is also restricted. The ratio of web depth to thickness is related to the grade of steel envisaged and whether stiffeners are provided. Minimum web thickness t_w for both unstiffened and vertically stiffened web plates are given in the following table:

Grade of steel	Unstiffened webs	Vertically stiffened webs
All	—	1/180 of the smaller clear panel dimensions
43	$d_1/85$	$d_2/200$
50	$d_1/75$	$d_2/180$
55	$d_1/65$	$d_2/155$

Note: This table relates to items 1 and 2 Clause 27f of BS 449. d_1 is the clear depth measured between the inside faces of the flanges; d_2 is twice the clear depth measured between the compression flange and the neutral axis.

On no account should the greater clear panel dimension of a web plate exceed $270 t_w$.

From the foregoing it would appear that the use of stiffeners is optional: this is not so. The table above is concerned with the problem of shear instability briefly referred to in the previous section on design of rolled beams on p. 31. Shear instability is not regarded as a problem if the ratio of web depth to thickness is less than the numerical factor given in the second column of this table, and "intermediate" stiffeners may then be omitted.

When the web thickness lies between the ratios given in the second and third columns intermediate stiffeners must be provided along the span of the girder at a spacing not in excess of $1 \cdot 5 d_1$ to comply with clause 28b(i). These stiffeners must in themselves be stiff enough to give restraining action to the web and for this purpose their moment of inertia about the longitudinal centre of the web should not be less than

$$I_s = \frac{d_1^3 t_w^3}{s^2} \times 1 \cdot 5 \times 10^{-4} \text{ cm}^4,$$

where s is the maximum permitted clear distance in millimetres between stiffeners for a web thickness t_w.

If an incoming beam is attached to an intermediate stiffener the inertia obtained above should be increased by an amount

$$I = \frac{150 M D^2}{E t_w} \text{ cm}^4$$

Similarly, if a lateral force is applied then I_s should be increased by

$$I = \frac{0 \cdot 3 P D^3}{E t_w} \text{ cm}^4,$$

where M is the applied moment in kilonewton metres, and P is the lateral force in kilonewtons.

The stiffeners themselves must not become unstable elements. For this reason the projection of plate stiffeners from the face of the web plate should not exceed twelve times their thickness unless the free edge is stiffened. Other stiffeners should not project more than $16t$, $14t$ and $12t$ for steel grades 43, 50 and 55 respectively, t being the thickness of the stiffener. Special stiffeners termed "load-bearing stiffeners" should be placed at points of support and at other positions within the span where concentrated loads would overstress the web. Thus those intermediate stiffeners carrying incoming beams would fall into this category.

Where at points of support torsional restraint is not provided by other means (see remarks in the previous section) then load-bearing stiffeners should be designed to give this restraint. Complying with clause 28a (iii) the minimum moment of inertia required of the stiffener is

$$I = \frac{D^3 T}{250} \cdot \frac{R}{W},$$

where W is the total load on the girder.

Working and permissible stresses

Calculating the actual (working) stresses due to bending, shear, etc. is no different from that of rolled beams, therefore no further mention need be made here.

The determination of permissible stresses is slightly more involved. Generally permissible stresses in plate girders are lower than for rolled beams. Now the design of plate girders is virtually a wide-open exercise since almost any depth, breadth or plate thickness may be used provided the detailed proportions are not violated.

It is a known fact that the yield strength of a given steel decreases as the material thickness increases and this is more evident with higher-strength steels. To achieve a good standard of design, plate thicknesses should be the lowest possible consistent with the foregoing remarks to produce the highest stresses in conformity with the specification.

Bending stresses

The tensile stress comes from Table 2 of B.S. 449, any reductions being due to material thickness. The compressive bending stress also comes from this table with similar reductions for material thickness, provided the compression flange is fully restrained. If not then lateral instability of this flange must be investigated in accordance with clause 20. Methods of obtaining end restraint and factors relating to effective length were discussed in the previous section and are applicable here. Partly or totally unrestrained compression flanges derive their permissible stress

p_{bc} from the critical stress C_s, which in turn depends upon the following factors:

1. the effective slenderness ratio l/r_y of the girder;
2. the ratio D/T, which in turn depends upon the degree of flange curtailment;
3. whether the flanges have equal moments of inertia about the yy-axis of the girder.

From (3) three cases obtain; these are:

Case I. Both flanges have equal moments of inertia about the yy-axis of the girder.

Case II. The inertia of compression flange about the yy-axis of the girder exceeds the tension flange.

Case III. The inertia of the tension flange about the yy-axis of the girder exceeds the compression flange.

Within these three cases it is assumed that I_{xx} of the girder exceeds I_{yy}; that the girder has a single web plate, and that the ratio of the flange thicknesses does not exceed three. Box girders and multiple-web girders are not covered by B.S. 449. Each of the above three cases will be discussed separately.

Case I

For the simplest type of plate girder in which the depth and width are constant throughout the span, the calculation of l/r_y and D/T is straightforward.

The critical compressive stress may be found from the expression:

$$C_s = A = \left(\frac{1675}{l/r_y}\right)^2 \sqrt{\left[1 + \frac{1}{20}\left(\frac{l/r_y}{D/T}\right)^2\right]}$$

or C_s may be obtained directly from Table 7 of B.S. 449 and converted to a permissible stress p_{bc} by reference to Table 8 of B.S. 449. However p_{bc} should not exceed Table 2 values.

If the flange plates are curtailed in width or thickness between points of lateral restraint the effect of this will, because the section is not quite so stiff, reduce the C_s value and the corresponding p_{bc} value. Hence D/T must be calculated taking into account the amount of curtailment. T now becomes the "effective" thickness of the compression flange at the position of the maximum bending moment.

$$\text{Effective thickness} = K_1 \times \text{mean thickness}$$

where K_1 depends upon the factor N which is the ratio of the total area of both flanges at the position of least bending moment to the corresponding area at the position of greatest bending moment between points of flange restraint.

The relationship between K_1 and N is shown graphically in Fig. 23. N should not be less than 0·25.

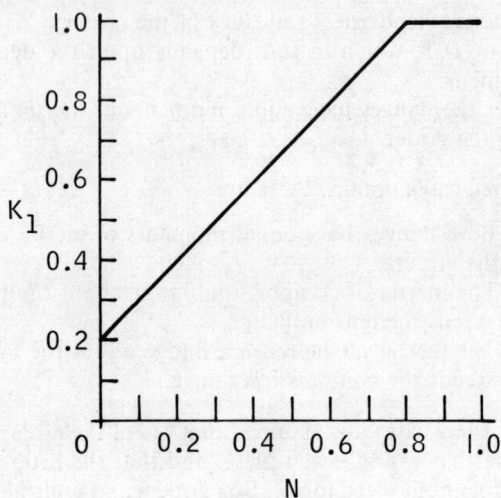

FIG. 23

Case II

When the compression and tension flanges have unequal I_{yy} values the shear centre (a term used in torsional analysis) no longer coincides with the position of the neutral axis for the whole cross section.

The position of the shear centre may be found from the expression

$$m = \frac{e}{h} = \frac{I_{yyc}}{I_{yyc} + I_{yyt}}$$

where I_{yyc} and I_{yyt} are the minor axis moments of inertia about the compression and tension flanges respectively.

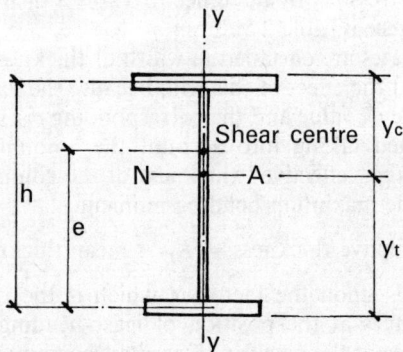

FIG. 24

For a Case II girder I_{yyc} exceeds I_{yyt} which means that the shear centre is above the neutral axis and m now exceeds 0·5. The effect of this is to improve the value of C_s because there is less torsional instability.

The critical compressive stress is obtained from

$$C_s = A = K_2 B$$

where $A = \left(\frac{1675}{l/r_y}\right)^2 \sqrt{\left[1 + \frac{1}{20}\left(\frac{l/r_y}{D/T}\right)^2\right]}$

and $B = \left(\frac{1675}{l/r_y}\right)^2$

The improvement in C_s is reflected in the term $K_2 B$ which is positive. K_2 is related to the ratio m as previously mentioned and is shown graphically in Fig. 25.

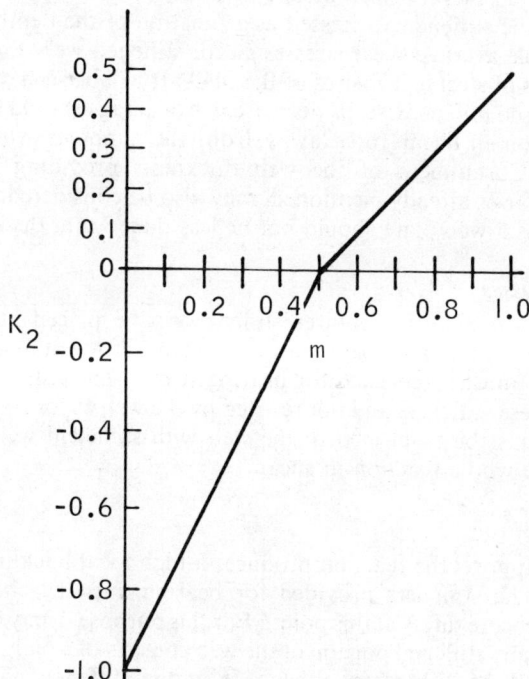

FIG. 25

Case III

Here the argument is similar to Case II except that the shear centre is below the neutral axis because I_{yyt} exceeds I_{yyc}, and m is less than 0·5. A less stable situation has now obtained and C_s will therefore decrease.

In this case $$C_3 = (A + K_2B)\frac{y_c}{y_t}$$

Both A and B have the meanings previously given.

Since m is less than 0.5 the value for K_2 will be negative (*see* Fig. 25). A further correction appears in the ratio y_c to y_t whose meaning will be obvious from Fig. 24.

Note: A detailed discussion of the critical stress equations may be found in the Proceedings of the ICE August 1956.

Shear stresses

If the d/t_w ratio of the web plate is low and stiffeners are not required (*see* table on p. 41 above) no reduction in permissible shear stress is suffered and Table 11 values in B.S. 449 may be used. When the d/t_w ratio is high the permissible shear stress is reduced to allow for instability of the web. The level of permissible shear stress also depends upon the spacing of the stiffeners expressed as a function of the depth, d.

Permissible average shear stresses p_q' for stiffened webs are given for three grades of steel in Table 12 of B.S. 449. It will be seen that as both ratios d/t_w and a/d increase, p_q' decreases. Advantage should be taken of the reduction in shear force away from the supports when spacing stiffeners. Curtailment of the web thickness, providing it remains within the ratios already mentioned, may also be considered. For practical reasons a web plate should not be less than 8 mm thick.

Bearing stresses

Since clause 28a (iii) requires stiffeners to be placed at points of support the contact area between the outstand of the stiffener and the flange plate must be checked for bearing stress. The entire reaction is taken by these stiffeners and not resisted by the web as for rolled beams. Stiffeners must be connected to the web with sufficient weld metal to transfer the whole reaction in shear.

Buckling stresses

At the supports the reaction produces a high local buckling stress in the web. The stiffeners provided for bearing must be checked as a column to ensure safety at this point. For this purpose it may be assumed that the locally stiffened portion of the web consists of a pair of stiffeners plus a length of web plate, either side of the stiffeners equal, where available, to $20t_w$. The radius of gyration should be taken parallel to the axis of the web, and the effective length of the stiffener taken at $0.7d$.

The actual buckling (compressive) stress must not exceed the appropriate value of p_c from Table 17 of B.S. 449.

Combined stresses

In accordance with clause 14d combined stresses should be checked at other points of concentrated load and stiffeners provided if found necessary.

Construction details

Modern plate girders are of all-welded construction. Attachment of the flanges to the web is usually by fillet welds which may be either continuous or intermittent. The minimum weld leg length should be considered as 6 mm. Intermittent fillet welds should preferably not be used if the girder is subject to fatigue loading.

Where the flanges are made up from different plate thicknesses to produce curtailment, the two parts should be full strength but welded together and the thicker plate chamfered down.

Very long plate girders will create problems in transportation and should therefore be delivered in two or more lengths and spliced at site: splicing may be by welding or bolting. Full-strength butt welds or friction-grip bolts with suitably proportioned cover plates should be used.

CRANE BEAMS

In industrial premises where overhead travelling cranes are employed a special beam termed a crane beam or gantry girder is provided. This beam or gantry has a bridge or flat bottom rail clipped to its compression flange which provides a continuous running surface for the crane. A crane beam is quite different from a floor beam in that the loading is almost totally live and therefore dynamic in character.

The building designer will know from the brief that a crane is required which has a specified lifting capacity. However, the crane beam cannot be designed until the loading details have been obtained from the crane supplier.

Information required by the crane supplier includes the following:

1. The crane capacity, *i.e.* lifted load;
2. its span centre to centre of gantry beams;
3. manually or electrically operated;
4. duty, *i.e.* infrequent, average or heavy use.

From these details the crane supplier will provide the following information:

1. the crane capacity (as specified);
2. weight of the crane;
3. weight of the crab;
4. minimum hook approach;
5. maximum anticipated wheel load (static);
6. spacing of end carriage wheels and clearance.

With this information the design loading applied to the gantry girder

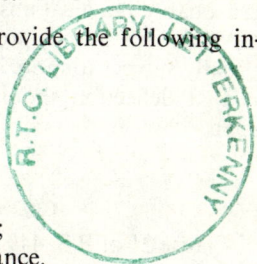

may now be computed. Because of the character of the loads to be resisted the designer makes allowance for the effects of impact, vibration, acceleration and retardation both longitudinally and transversally, and shock from slipping of slings. These allowances depend upon whether the crane is manually or electrically operated.

Unless there are special circumstances the following allowances are employed:

1. Loads acting vertically through end carriage wheels are increased
   by 25 per cent for electric overhead cranes
   10 per cent for hand-operated cranes
2. Loads acting horizontally and transverse to the gantry beam due
   to the combined weight of the crab and lifted load
   10 per cent for electric overhead cranes
   5 per cent for hand-operated cranes
3. Loads acting horizontally and parallel to the gantry beam due to
   the static vertical wheel loads
   5 per cent for both electric and hand-operated cranes.

Usually the forces obtained from Item 1 combined with Item 2 control the design, and in this respect the permissible stresses may be increased by 10 per cent. Since crane beams may be subject to fatigue, this should be allowed for where appropriate and reference should be made to the British Standard on steel girder bridges (previously B.S. 153, now B.S. 5400).

A crane beam usually consists of a universal beam with a channel welded to the compression flange. Additional flange plates may also be provided where appropriate. Details relating to the composition, dimensions and section properties are given in the CONSTRADO Structural Steelwork Handbook, where they are specifically referred to as gantry girders.

As this type of beam carries both vertical and horizontal transverse loading it presents problems of torsion, however, practice has established a simple approach. In effect, the vertical loading is resisted by the compound section whilst the transverse loading is assumed to be resisted by the compression flange only. This simple technique appears to give reasonably practical answers.

Whilst no special provision is made in B.S. 449, crane beams are usually subject to more stringent deflection rules. The accepted practice is that deflection should not exceed one five-hundredth part of the span.

CONCRETE ENCASED BEAMS

Clause 21 of B.S. 449 allows the concrete encasement to a beam to be taken into account in assessing the permissible bending stress, but the advantage of improved stress only obtains when the beam is laterally unsupported. The provisions of this clause do not lead to a truly composite design.

The effect of the concrete encasement, which should give a cover of at least 50 mm and have a 28-day strength of 21 N/mm² is to increase the radius of gyration from the value given in the beam tables to 0·2 (B + 100). Thus the slenderness ratio decreases, giving an improved permissible bending stress which may not exceed 1·5 times the Table 3 value for the unencased beam, but with an upper limit taken from Table 2.

Similarly, beams termed "filler joists" may be designed in accordance with clause 29, and since they will be fully restrained higher bending stresses are permitted. The method discussed allows only grade 43 steel to be used with a concrete strength not less than 21 N/mm² at 28 days. Permissible bending stresses are increased depending upon the amount of cover given to the compression flange and the actual stresses calculated as plain beams. Spacing of filler joists is related to the cover thickness and the imposed loading. In effect this clause recognises the benefits of structural interaction and goes on to suggest that as an alternative the fillet joist and slab may be designed as a composite section but no guidance is given.

EXAMPLE 14

Figure 26 (b) shows layout of the a beam grillage which supports heavy industrial plant. The areas shown crossed carry a superimposed load of 5·0 kN/m² plus the self-weight of an open-grill floor mesh estimated at 1·0 kN/m². Positions and magnitude of plants are given on the plan. Design the grillage using grades 43 and 50 steel as appropriate.

SOLUTION

The approach to the design follows a logical progression; the order is:

1. Design common beams marked (12), (14), (16), (21), (23) and (25).
2. Design edge beams marked (11), (21), (17) and (27) all common.
3. Design plant beams marked (13), (23), (15) and (25) all common.
4. Design main beams marked (10), (20) and (30) all common.

Beams marked (12), (14), etc.

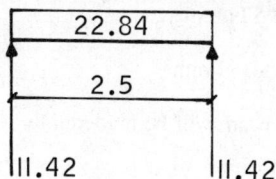

FIG. 26(a)

Loading		
Imposed	5 × 2·5 × 1·5 =	18·75
Grill	1 × 2·5 × 1·5 =	3·75
S.W., say		0·34
Total		22·84 kN

$$M = \frac{22\cdot84 \times 2\cdot5}{8} = 7\cdot14 \text{ kNm}$$

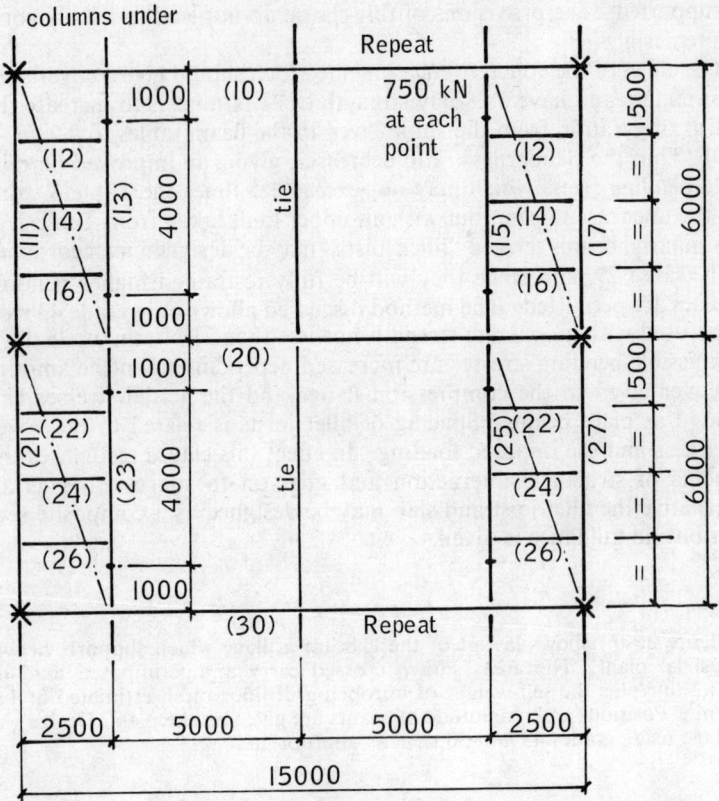

columns under

FIG. 26(b)

The beam is laterally unsupported on account of grill floor, and ends are located by web cleats. Take the effective length as 2·5 m (*see* fig. 26(a)).

Try 127 × 76 × 13·36 joist: $Z = 74·94 \times 10^3$ mm³, $r_y = 17·2$ mm, $D/T = 16·7$; use grade 43 steel.

$$\frac{l}{r} = \frac{2·5 \times 10^3}{17·2} = 146$$

From Table 3a by interpolation $p_{bc} = 138·5$ N/mm²,

$$f_{bc} = \frac{7·14 \times 10^6}{74·94 \times 10^3} = 95·3 \text{ N/mm}^2$$

Although this stress is low the next smaller beam will be inadequate. Check for deflection due to imposed load

$$\delta_p \ngtr \frac{L}{360} = \frac{2·5 \times 10^3}{360} = 6·95 \text{ mm,}$$

$$\delta_a = \frac{5WL^3}{384EI} = \frac{5 \times 18·75 \times 2·5^3 \, 10^6}{384 \times 210 \times 10^3 \times 475·9 \, 10^4} = 3·82 \text{ mm} < \delta_p$$

or, using the expression given in the text,

actual $\qquad \dfrac{L}{D} = \dfrac{2 \cdot 5 \times 10^3}{127} = 19 \cdot 7$

permissible $\qquad \dfrac{L}{D} = \dfrac{2800}{kf}$,

$$k = \dfrac{18 \cdot 75}{22 \cdot 84} = 0 \cdot 82 \qquad f = 95 \cdot 3,$$

$$\therefore \dfrac{L}{D} = \dfrac{2800}{0 \cdot 82 \times 95 \cdot 3} = 35 \cdot 9 > \text{actual}, \ \therefore \text{satisfactory}.$$

Calculations for web shear, bearing and buckling will show the section is more than adequate. Use $127 \times 76 \times 13 \cdot 36$ joist—grade 43.

Beams marked (11), (21), etc. (*see* Fig. 26 (c)):

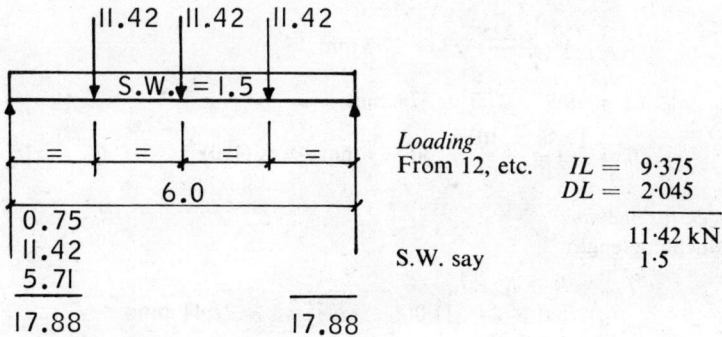

Fig. 26(c)

$$M = (17 \cdot 88 \times 3) - (0 \cdot 75 \times 1 \cdot 5) - (11 \cdot 42 \times 1 \cdot 5) = 35 \cdot 38 \text{ kNm}$$

The unsupported flange length is $1 \cdot 5$ m and is not critical, but calculation will be taken in fullness to illustrate.
Try $254 \times 102 \times 25$ U.B.: $Z = 265 \cdot 2 \times 10^3 \text{ mm}^3$, $r_y = 21 \cdot 4$ mm, $D/T = 30 \cdot 8$.

$$\dfrac{l}{r} = \dfrac{1 \cdot 5 \times 10^3}{21 \cdot 4} = 70$$

From Table 3

$$\text{for grade 43} \quad p_{bc} = 165 \text{ N/mm}^2,$$
$$\text{grade 50} \quad p_{bc} = 230 \text{ N/mm}^2,$$

$$f_{bc} = \dfrac{35 \cdot 38 \times 10^6}{265 \cdot 2 \times 10^3} = 133 \cdot 4 \text{ N/mm}^2$$

Checking for deflection and using approximate method as loading tends to a U.D. L

actual $\qquad \dfrac{L}{D} = \dfrac{6 \times 10^3}{257} = 23 \cdot 3,$

$$k = \frac{3 \times 9 \cdot 375}{3 \times 11 \cdot 42 + 1 \cdot 5} = 0 \cdot 79,$$

permissible
$$\frac{L}{D} = \frac{2800}{0 \cdot 79 \times 133 \cdot 4} = 26 \cdot 2,$$

check for shear

$$f_q' = \frac{17 \cdot 88 \times 10^3}{257 \times 6 \cdot 1} = 11 \cdot 4 \; N/mm^2 < p_q' \; (100)$$

Check for bearing

Assume beam carried on simple brackets cut from $90 \times 90 \times 8$ angles, root radius $8 \cdot 4$ mm. Bearing length $L_b = l_1 + l_2 + l_3$

$$l^1 = 8 \cdot 0 \; [\sqrt{(3)} + 1] + 11 \cdot 0 \; [\sqrt{(3)} - 1] - 5 = 29 \cdot 9 \; mm$$
$$l^2 = 0$$

$$l_3 = \frac{257 - 225}{2} \sqrt{3} = 27 \cdot 7 \; mm$$

$$\therefore L_b = 29 \cdot 9 + 27 \cdot 7 = 57 \cdot 6 \; mm$$

$$f_b = \frac{17 \cdot 88 \times 10^3}{57 \cdot 6 \times 6 \cdot 1} = 50 \cdot 8 \; N/mm^2 < p_b \; (190)$$

Check for buckling

Buckling length

$$L_{bu} = l_1 + l_2 + l_3,$$
$$l_1 = 8 \cdot 0 \times 2 + 11 \cdot 0 (2 - \sqrt{2}) - 5 = 22 \cdot 44 \; mm,$$
$$l_2 = 0,$$

$$l_3 = \frac{257}{2} = 128 \cdot 5 \; mm,$$

$$\therefore L_{bu} = 22 \cdot 44 + 128 \cdot 5 = 150 \cdot 94 \; mm,$$

$$f_{bu} = \frac{17 \cdot 88 \times 10^3}{150 \cdot 94 \times 6 \cdot 1} = 19 \cdot 4 \; N/mm^2,$$

$$\text{web } \frac{l}{r} = \frac{d_3}{t_w} \sqrt{3} = \frac{225 \times \sqrt{3}}{6 \cdot 1} = 64$$

From Table 17a

$$p_c = 122 \; N/mm^2$$

All stress conditions are shown to be satisfactory. As this bending stress is low and deflection is approaching limit use $254 \times 102 \times 25$ U.B. in grade 43.

Beams marked (13),(23), etc. (*see* Fig. 26(*d*)):

$M = 771 \cdot 33 \times 3 =$		2313·99
less	$750 \times 2 = 1500$	
	$11 \cdot 42 \times 1 \cdot 5 =$	17·3	
	$4 \cdot 2 \times 1 \cdot 5 =$	6·3	1523·43
$\therefore 2313 \cdot 99 - 1523 \cdot 43 =$			790·56 kNm

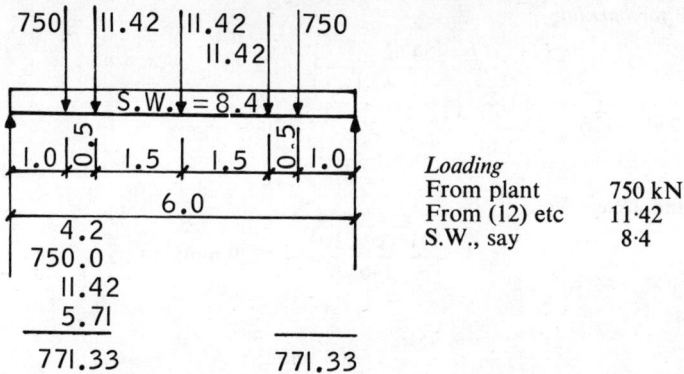

FIG. 26(d)

Loading

From plant	750 kN
From (12) etc	11·42
S.W., say	8·4

As beam is restrained at 1·5 m intervals take full stress and assume grade 50 steel

$$\text{required } Z = \frac{790·56 \times 10^6}{230} = 3437 \times 10^3 \text{ mm}^3$$

A 610 × 229 × 140 U.B. has $Z = 3626 \times 10^3$ mm³, giving $f_{bc} = 218$ N/mm².

Check for deflection

This is unnecessary as the actual L/D ratio is less than 10. (Generally if $L/D < 16$ deflection calculations may be ignored.)

Check for shear

$$f'_q = \frac{771·33 \times 10^3}{617 \times 13·1} = 95·6 \text{ N/mm}^2 < p'_q (140)$$

Check for bearing

$$p_b = 260 \text{ N/mm}^2 \quad \text{(Table 9),}$$

Required area

$$A_b = \frac{771·33 \times 10^3}{260} = 2966 \text{ mm}^2,$$

so that

$$L_b = \frac{2966}{13·1} = 226·4 \text{ mm}$$

Now the l_2 component $= 0$ and

$$l_3 = \frac{617 - 547·2}{2} \sqrt{3} = 60·4 \text{ mm}$$

then the stiff bearing length required is $l_1 = 226·4 - 60·4 = 166$ mm. This is outside the range offered by any simple angle bracket: defer until later.

Check for buckling

Web $\dfrac{l}{r} = \dfrac{547 \cdot 2 \times \sqrt{3}}{13 \cdot 1} = 72 \cdot 35$

from Table 17b

$$p_c = 146 \text{ N/mm}^2,$$

required area

$$A_{bu} = \dfrac{771 \cdot 33 \times 10^3}{146} = 5280 \text{ mm}^2$$

so that

$$L_{bu} = \dfrac{5280}{13 \cdot 1} = 404 \text{ mm}$$

Now the l_2 component is zero and

$$l_3 = \dfrac{617}{2} = 308 \cdot 5 \text{ mm}$$

then the stiff buckling length required is

$$l_1 = 404 - 308 \cdot 5 = 95 \cdot 5 \text{ mm}.$$

Clearly both bearing and buckling are special detail considerations. Two solutions are available. These are:

1. provide a built-up bracket system on the receiving beam to give a stiff length of about 170 mm; or
2. design the end connections to transfer the reaction to the receiving beam in shear only, thus eliminating bearing and buckling.

The final decision will depend upon the design of the receiving beam.
Beams marked (10), (20), (30): (*see* Fig. 26(*e*))

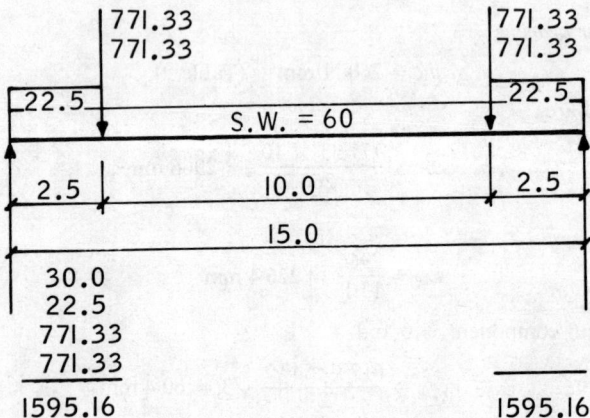

FIG. 26(*e*)

Loading

From (13), (23), etc.		771·33 kN
Local imposed as (12)		18·75
Local grill as (12)		3·75
S.W., say		60·00

$$M = 1595 \cdot 16 \times 7 \cdot 5 = \qquad 11\,963 \cdot 7$$

$$
\begin{array}{lrr}
\text{less} & 22 \cdot 5 \ \times 6 \cdot 25 = & 140 \cdot 6 \\
 & 30 \cdot 0 \ \times 3 \cdot 75 = & 112 \cdot 5 \\
 & 1542 \cdot 66 \times 5 \cdot 00 = & 7713 \cdot 3 \qquad 7966 \cdot 4 \\
 & & \overline{3997 \cdot 3 \ \text{kNm}}
\end{array}
$$

required $Z = \dfrac{3997 \cdot 3 \times 10^6}{230} = 17\,380 \times 10^3 \ \text{mm}^3$

This is outside the range of U.B.s; examine a plate girder. Preliminary calculations for approximate size only.

Try a plate girder about 1·2 m deep, put $f_{\text{bc}} = 215 \ \text{N/mm}^2$, $f'_{\text{q}} = 120 \ \text{N/mm}^2$. The area of one flange can be shown to be approximately:

$$\text{flange area} \ A_{\text{f}} = \frac{6M}{7Df_{\text{bc}}}$$

$$= \frac{6 \times 3997 \cdot 3 \times 10^3}{7 \times 1 \cdot 2 \times 215} = 13\,300 \ \text{mm}^2$$

say 450×30 flange. $\qquad A_{\text{f}} = 13\,500 \ \text{mm}^2$

web area $\qquad\qquad A_{\text{w}} = \dfrac{1595 \cdot 16 \times 10^3}{120} = 13\,300 \ \text{mm}^2$

say 1200×12 web. $\qquad A_{\text{w}} = 14\,400 \ \text{mm}^2$

Hence this section is roughly in the proportions suggested in the text. This section will be examined in detail as a Case 1 girder. The properties are:

$A = 2 \times 13\,500 + 14\,400 = 41\,400 \ \text{mm}^2$

$I_{xx} = \left[\dfrac{450 \times 30^3}{12} + (450 \times 30 \times 615^2) \right] 2 + \dfrac{12 \times 1200^3}{12}$

$\qquad = 11\,942 \times 10^6 \ \text{mm}^4$

$Z_{xx} = \dfrac{11\,942 \times 10^6}{630} = 18\,955 \times 10^3 \ \text{mm}^3$

$I_{yy} = \dfrac{2 \times 30 \times 450^3}{12} + \dfrac{1200 \times 12^3}{12} = 456 \times 10^6 \ \text{mm}^4,$

$r_{yy} = \sqrt{\left(\dfrac{456 \times 10^6}{414 \times 10^2} \right)} = 105 \ \text{mm},$

$\dfrac{D}{T} = \dfrac{1260}{30} = 42$

Now $T/t_{\text{w}} = 30/12 = 2 \cdot 5 > 2$, and $d/t_{\text{w}} = 1200/12 = 100 > 75$; then the 20 per cent increase in C_{s} does not apply and

$$C_{\text{s}} = \left(\frac{1675}{l/r_y} \right) \bigg/ \left[1 + \frac{1}{20} \left(\frac{l/r_y}{D/T} \right)^2 \right],$$

where

$$\frac{l}{r_y} = \frac{5000}{105} = 47 \cdot 6,$$

$$\therefore \quad C_s = \left(\frac{1675}{47 \cdot 6}\right)^2 \sqrt{\left[1 + \frac{1}{20}\left(\frac{47 \cdot 6}{42}\right)^2\right]} = 1262 \text{ N/mm}^2$$

from Table 8 for $C_s = 1262$, $p_{bc} = 223 \cdot 5$ N/mm², but from table 3 $p_{bc} = 215$ N/mm² (use this value) and

$$f_{bc} = \frac{3997 \cdot 3 \times 10^6}{18\,955 \times 10^3} = 211 \text{ N/mm}^2$$

which is satisfactory for bending.

Check for shear

$$f_q' = \frac{1595 \cdot 16 \times 10^3}{1200 \times 12} = 92 \cdot 3 \text{ N/mm}^2 < p_q'(140),$$

but p_q' depends upon the spacing of the web stiffeners which must be provided since $d/t_w > 75$.

Between the support and incoming beam (13) the clear distance is $2 \cdot 5$ m which is greater than $1 \cdot 5d$. Provide two pairs of stiffeners at the support, say 300 mm, apart, and a further two pairs 400 mm apart at the position of beam (13), then subdivide the reduced distance, now 2150 mm equally, making the spacing of the intermediate stiffeners 1075 mm. Then spacing factor is

$$\frac{1075}{1200} = 0 \cdot 9 \text{ (approx.)}$$

and p_q' from Table 12b is 139 N/mm², hence web plate is satisfactory.

Web stiffeners

1. At supports
Load to be resisted in bearing is:

from plate girder	1595·16
from beam (11)	17·88
from beam (21)	17·88
	1630·92 (say 1631 kN)

using four stiffeners bearing area required at $p_b = 260$ N/mm²

$$A_b = \frac{1631 \times 10^3}{4 \times 260} = 1568 \text{ mm}^2$$

Make stiffeners 140×15 to comply with clause 28b(iii) and allow a 25 mm chamfer to clear web flange welds; area provided is $(140 - 25)15 = 1725$ mm². Check this stiffener arrangement as a column (*see* Fig. 26(f)).

Area available for column action is:

gross stiffener area	$4 \times 140 \times 15$	8400
plus web area	640×12	7680
		16 080 mm²

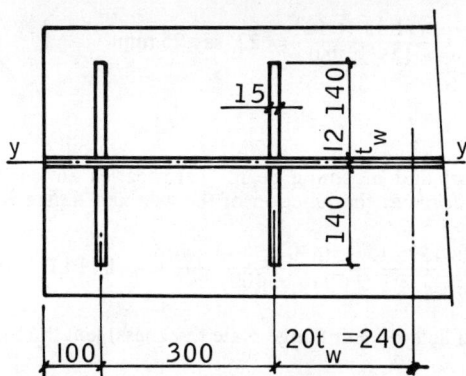

FIG. 26(f)

$$I_{yy} = \frac{30 \times 292^3}{12} + (640 - 30)\frac{12^3}{12} = 62\,329 \times 10^3 \text{ mm}^4,$$

$$r_y = \sqrt{\left[\frac{62\,329 \times 10^3}{16\,080}\right]} = 62 \text{ mm},$$

$$\frac{l}{r} = \frac{0.7 \times 1200}{62} = 13.6,$$

giving $p_c = 208$ N/mm^2;

$$f_c = \frac{1631 \times 10^3}{16\,080} = 101.2 \text{ N/mm}^2$$

This stiffener arrangement is satisfactory.

2. Intermediate stiffeners
These are non-load-bearing, but their moment of inertia must be not less than

$$I = 1.5 \times 10^{-4} \times \frac{d^3 t^3_w}{S^2} \text{ cm}^4$$

$$= 1.5 \times \frac{10^{-4} \times 1200^3 \times 12^3}{(1.5 \times 1200)^2} = 138.2 \text{ cm}^4 \quad \text{(or } 1382 \times 10^3 \text{ mm}^4\text{)}.$$

Provide 100×10 stiffeners either side of web.

3. At incoming beams (13), (23)
   These stiffeners have been arranged so that they do not actually transmit the beam reaction, they may therefore be regarded as intermediate. However the stiffened bracket must be designed.
   Make the table of the bracket 25 mm thick with an outstand of 170 mm to suit bearing requirements and the width to suit the spacing of the stiffeners, i.e. $400 - 12 = 382$ mm.
   The table needs to be stiffened and the bearing area required, allowing for an outstand of 160 mm less 25 mm chamfer, gives a thickness of

$$t = \frac{771 \cdot 33 \times 10^3}{135 \times 260} = 22, \text{ say 25 mm}$$

Welding

1. Web to flanges

Between the support and incoming beam (13) shear is almost constant. The horizontal shear force at the junction of the web and flange is

$$q_b = \frac{QA\bar{Y}}{I} = \frac{1595 \cdot 16 \times (450 \times 30) \times 615}{11\,942 \times 10^6} = 1 \cdot 11 \text{ kN/mm}$$

using a pair of 8 mm fillets (to suit flange plate thickness) length required per metre is

$$\frac{1 \cdot 11 \times 10^3 \times 1000}{2} = 0 \cdot 7 \times 8 \times 160l$$

giving

$$l = \frac{1 \cdot 11 \times 10^6}{2 \times 0 \cdot 7 \times 8 \times 160} = 620 \text{ mm}$$

arrange intermittent welding to suit this length, but make weld continuous for at least 500 mm at support and at position of beam (13).

2. End-bearing stiffenerers

The total reaction is 1631 kN, and with four stiffeners, each with two weld runs of not less than 5 mm to suit plate thickness, length required is

$$\frac{1631 \times 10^3}{4 \times 2} = 0 \cdot 7 \times 5 \times 160l$$

giving

$$l = \frac{1631 \times 10^3}{4 \times 2 \times 0 \cdot 7 \times 5 \times 160} = 364 \text{ mm}$$

Provide intermittent fillet welds to suit this length. Ends of stiffeners must be fitted tight to flanges.

3. Stiffened brackets

Load to be resisted is 771·3 kN. Take stiffener down to contact lower flanges. Then length available for welding is

$$l = 1230 - (617 + 25 + 30) = 558 \text{ mm}$$

Using two welds, their size is

$$\frac{771 \cdot 33 \times 10^3}{2} = 0 \cdot 7s \times 160 \times 558$$

giving

$$s = \frac{771 \cdot 33 \times 10^3}{2 \times 0 \cdot 7 \times 160 \times 558} = 6 \cdot 2 \text{ say 8 mm}$$

Typical details are shown in Fig. 27.

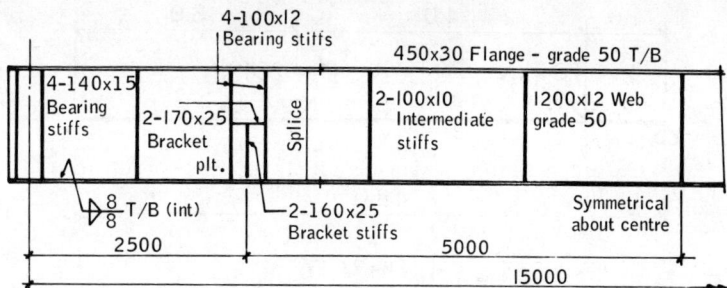

FIG. 27

EXAMPLE 15

Design a simply supported crane gantry girder to span 8 m using grade 43 steel. Crane details are as follows:

Crane capacity (lifted load)	300 kN
Weight of crane excluding crab	250 kN
Weight of crab	50 kN
Minimum hook approach	1 m
Span of crane between gantries	17 m
Centres of end carriage wheels (two wheels per carriage)	4 m

SOLUTION

Since the maximum static wheel load is not given it may be calculated thus:

from weight of crane on 4 wheels $\frac{250}{4} = 62 \cdot 5$ kN

from crab at min. approach $\frac{16}{17} \times \frac{50}{2} = 23 \cdot 5$

from load as above $\frac{16}{17} \times \frac{300}{2} = \underline{141 \cdot 2}$

Maximum static load per wheel $= 227 \cdot 2$ kN

For an E.O.T. crane allow 25 per cent impact *etc*, therefore max wheel load

$$= 1 \cdot 25 \times 227 \cdot 2 = 284 \text{ kN}$$

for horizontal transverse surge allow 10 per cent lifted load + crab

$$= \frac{0 \cdot 1}{4} (300 + 50) = 8 \cdot 75 \text{ kN}$$

For horizontal longitudinal surge allow 5 per cent of wheel loads

$$= 0 \cdot 05 \times 227 \cdot 2 \times 2 = 22 \cdot 7 \text{ kN}$$

Design for bending

From a knowledge of influence lines it can easily be shown that the maximum bending moment will be obtained when one wheel and the centroid of both wheels are placed equidistant about the centre of the span. This position is shown in Fig. 28.

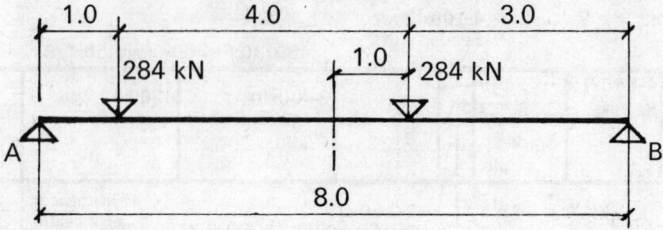

Fig. 28

For this condition the bending is maximum at the 284 kN load nearest the span centre.

Taking moments about A for R_B (vertical)

$$8R_B - 1 \times 284 + 5 \times 284 = 0$$

from which $$R_B = 213 \text{ kN}$$

Maximum moment $= 213 \times 3 = 639$ kN m

Similarly for the horizontal transverse surge, the wheels will be in the same position.

Taking moments about A for R_B (horizontal)

$$8R_B + 1 \times 8.75 + 5 \times 8.75 = 0$$

from which $$R_B = 6.56 \text{ kN}$$

Maximum moment $= 6.56 \times 3 = 19.69$ kN m

Since the beam size cannot be obtained direct, the following *preliminary* assumptions will be made:

1. Beam is fully restrained allowing max p_{bc};
2. Vertical bending is resisted on U.B. using p_{bc};
3. Horizontal bending is resisted on channel using $0.25\ p_{bc}$.

Ignoring self weight at this stage

$$\text{vertical } Z = \frac{639 \times 10^3}{165} = 3873 \text{ cm}^3$$

$$\text{horizontal } Z = \frac{19.69 \times 10^3}{0.25 \times 165} = 477 \text{ cm}^3$$

∴ from section tables try a gantry girder consisting of a $610 \times 229 \times 140$ U.B. plus $381 \times 102 \times 55$ channel.

The properties for the combined section are as follows:

top flange, $\qquad\qquad Z_{xx} = 6606 \text{ cm}^3 \qquad r_y = 8.84 \text{ cm}$

bottom flange, $\qquad\quad Z_{xx} = 3982 \text{ cm}^3 \qquad \dfrac{D}{T} = 26.4$

top flange, $\qquad\qquad Z_{yy} = \ 900 \text{ cm}^3$

Final design

1. Bending

Self weight $= 140$ (U.B.) $+ 55$ (channel) $+ 35$ (rail) $= 230$ kg/m say 235 kg/m to allow for welding and other fittings.

Reactions are 9·4 kN each (by inspection)

At 3 m from B selfweight $M = 9·4 \times 3 - \dfrac{2·35 \times 3^2}{2} = 17·6$ kN m

∴ total bending moment $= 639 + 17·6 = 656·6$ kN m

vertical f_{bc} $= \dfrac{656·6 \times 10^6}{6606 \times 10^3} = 99·4$ N/mm²

horizontal f_{bc} $= \dfrac{19·69 \times 10^6}{900 \times 10^3} = 21·9$ N/mm²

Hence Σf_{bc} $= 99·4 + 21·9 = 121·3$ N/mm²

vertical f_{bt} $= \dfrac{656·6 \times 10^6}{3982 \times 10^3} = 164·9$ N/mm²

As there is no lateral restraint to compression flange, take effective length as 8 m, then,

$$\frac{l}{r_y} = \frac{8000}{88·4} = 90·5$$

From Table 3 (B.S. 449) for $l/r = 90·5$ and $D/T = 26·4$, $p_{bc} = 165$ N/mm². Since $f_{bc} < 1·1 \, p_{bc}$ and $f_{bt} < p_{bt}$. (165), section is satisfactory for bending.

2. Shear
Again, from influence lines, the maximum shear occurs when one wheel is over a support, then for vertical loads

$$\text{maximum } R_A = 9·4 + 284 + \frac{284}{2} = 435·4 \text{ kN}$$

the actual depth of the U.B. is 617 mm and the web thickness is 13·1 mm so that

$$\text{shear stress } f_s = \frac{435·4 \times 10^3}{617 \times 13·1} = 53·9 \text{ N/mm}^2$$

Since this is less than $p'_q = 100$, section is satisfactory.
For the horizontally transverse loads

$$\text{maximum } R_A = 8·75 + \frac{8·75}{2} = 13·13 \text{ kN}$$

Taking this entirely on the channel web, $d = 381$, $t = 10·4$

$$\text{shear stress } f_s = \frac{13·13 \times 10^3}{381 \times 10·4} = 3·3 \text{ N/mm}^2$$

which is very small.

3. Welding between U.B. and channel

This involves the use of the expression $qb = \dfrac{QA\bar{Y}}{I}$

For vertical shear $Q = 435\cdot4$ kN, $A = 70\cdot19$ cm² which is the area of the channel, $\bar{Y} = 21\cdot1$ cm the distance from the neutral axis to the centroid of the channel and $I = 155\ 882$ cm⁴

$$\text{vertical } qb = \frac{435\cdot4 \times 10^3 \times 70\cdot19 \times 10^2 \times 21\cdot1 \times 10^1}{155\ 882 \times 10^4} = 413\cdot7 \text{ N/mm}$$

This shear is being resisted by two welds, so that shear per weld equals 206·9 N/mm.

For horizontal transverse shear,

$Q = 13\cdot13$ kN, and A = area of channel beyond tip of U.B. flange
$\quad\quad\quad\quad\quad$ Y = distance from vertical axis of the beam to this area
$\quad\quad\quad\quad\quad$ I = inertia of top flange only = 17 148 cm⁴

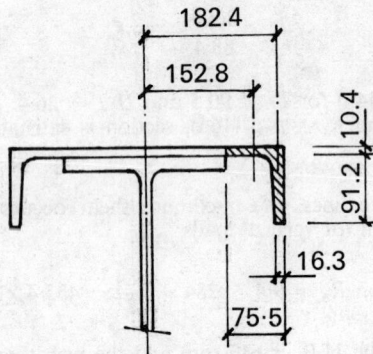

FIG. 29

From the figure, $A\bar{Y} = a_1y_1 + a_2y_2$
$$75\cdot5 \times 10\cdot4 \times 152\cdot8 \quad 120 \times 10^3$$
$$+ 91\cdot2 \times 16\cdot3 \times 182\cdot4 \quad 271 \times 10^3$$
$$\overline{391 \times 10^3}$$

Hence $\quad\quad\quad qb = \dfrac{13\cdot13 \times 10^3 \times 391 \times 10^3}{17\ 148 \times 10^4} = 29\cdot9 \text{ N/mm}$

now this is the shear being resisted by one weld.
Total weld shear $\quad\quad\quad = 206\cdot9 + 29\cdot9 = 236\cdot8$ N/mm
for mild steel fillet welds, $p_w = 115$ N/mm²
so that weld leg length is

$$l = \frac{236\cdot9}{0\cdot7 \times 115} = 2\cdot94 \text{ mm}$$

Use 6 mm F.W.

EXAMPLE 16

The two-span beam shown in Fig. 30 is unsupported between the applied loads and the reactions, and is to be fully concrete encased. Design the beam in grade 43 steel.

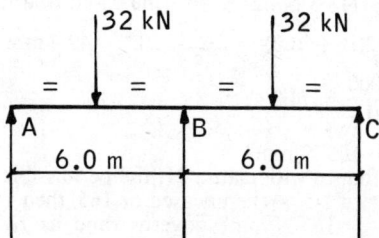

FIG. 30

SOLUTION

$$R_{AB} = R_{CB} = \frac{5 \times 32}{16} = 10 \text{ kN}$$

$$R_{BA} = R_{BC} = \frac{11 \times 32}{16} = 22 \text{ kN}$$

$$M_{BA} = \frac{3WL}{16} = \frac{3 \times 32 \times 6}{16} = 36 \text{ kNm}$$

$$M_D = \frac{5WL}{32} = \frac{5 \times 32 \times 6}{32} = 30 \text{ kNm}$$

Try a $254 \times 102 \times 28$ U.B. $Z = 307 \cdot 6 \times 10^3$ mm^3, $D/T = 26$, $r_y = 21 \cdot 9$ mm. Giving 50 mm concrete cover all-round overall dimensions are $B_c = 202$, $D_c = 360$.

Cross-sectional area of encasement = $202 \times 360 =$ 72 720 mm^2
Less that of beam   3620

Total   69 100 mm^2

then taking "weight" of concrete at 24 kN/m^3 then self-weight of concrete plus beam is

$$W = \frac{69\ 100 \times 24}{10^6} + 0 \cdot 28 = 1 \cdot 94 \text{ kN/m}$$

Additional moment at B is

$$M_{BA} = \frac{1 \cdot 94 \times 6^2}{8} = 8 \cdot 74 \text{ kNm},$$

$\therefore$ maximum moment to be resisted is

$$M = 36 + 8 \cdot 74 = 44 \cdot 74 \text{ kNm}$$

giving a stress on the plain section of

$$f_{bc} = \frac{44 \cdot 74 \times 10^6}{307 \cdot 6 \times 10^3} = 145 \cdot 3 \text{ N/mm}^2$$

For the uncased beam:

$$\frac{l}{r} = \frac{3000}{21\cdot9} = 137,$$

from Table 3a $p_{bc} = 118\cdot3$ N/mm². For the cased beam:

$$r = 0\cdot2(B + 100) = 0\cdot2 \times 202 = 40\cdot4 \text{ mm},$$

$$\frac{l}{r} = \frac{3000}{40\cdot4} = 74\cdot3,$$

from Table 3a $p_{bc} = 165$ N/mm²

However, in accordance with clause 21, the permissible stress on the cased section must not exceed $1\cdot5 \times$ the uncased or 165, then $1\cdot5 \times 118\cdot3 = 177\cdot4$ N/mm². Hence $p_{bc} = 165$ N/mm² governs, and as $f_{bc} < p_{bc}$ the section selected is satisfactory.

Chapter 3

COLUMNS AND TIES

The term column refers to any structural member working in compression. B.S. 449 uses the term "compression members" and talks of struts, stanchions and columns. It is probably better to think of a column or stanchion as a vertical support in a building and a strut as a compression member in a truss.

A tie infers a tension member and this is how it is meant in B.S. 449 and in this chapter. A tie is sometimes an unloaded beam connected between columns and serves to reduce the effective length of that member.

DESIGN OF TENSION MEMBERS

Tension members may be either concentrically loaded or eccentrically loaded, depending how the member is arranged in relation to its terminal connections. Ties may be fabricated from angles, tees or channels and sometimes beams or column sections. The last named should be avoided, particularly when the loading results in heavy, thick members.

Concentrically loaded ties

If the terminal connections are formed by welding and no holing is anywhere required the least area to resist the applied load is merely

$$A = \frac{F}{p_t},$$

and a section or sections selected to meet this.

When the terminal connections are formed by bolting the effect of the holes must be taken into account. The net area (because there are holes) to resist the applied load is

$$A_n = \frac{F}{p_t}$$

If the thickness of metal is t and the hole diameter d and n number are deducted from any cross-section the gross area required is

$$A_g = A_n + ndt,$$

and a section is selected to give this area, which of course should have the same thickness t.

Eccentrically loaded ties

A single angle connected through one leg will either have a non-uniform tensile stress in the unconnected leg or this leg will be partly in compression. B.S. 449 gives simplified rules for dealing with this case.

Calculate the net area required, *i.e.* $A_n = F/p_t$ and select a section giving an area at least 25 per cent greater than A_n. Using this section let A_1 be the net area of the connected leg and A_2 be the gross area of the outstanding or unconnected leg. This second area is reduced to allow for the non-uniformity of stress by a factor k. The actual net area then becomes

$$A'_n = A_1 + kA_2,$$

where

$$k = \frac{3A_1}{3A_1 + A_2},$$

and the section will be adequate providing

$$A'_n \geqslant A_n$$

A pair of angles or a single tee attached to one side of a gusset will behave in a similar manner then the reduction factor takes the value

$$k = \frac{5A_1}{5A_1 + A_2}$$

Ties formed from single channels do not comply with the simplified rules and should be designed as eccentrically loaded members. Both the direct stress and the stress due to bending must be computed. This means that a member must almost be selected at random and checked for suitability. As a rough guide select a member having an area about twice that required for direct stress only. The actual stresses will be

$$f_t = \frac{F}{A_n}$$

and

$$f_{bt} = \frac{Fe}{Z}$$

which must be combined in the form

$$\frac{f_t}{p_t} + \frac{f_{bt}}{p_{bt}} \not> 1$$

Both concentric and eccentrically loaded ties should be proportioned such that the allowable tensile stress p_t given in Table 19 of B.S. 449 is not exceeded. Tensile bending stresses should not exceed Table 2 values.

Construction details

Staggered holing may present additional calculation in relation to net areas and this is dealt with in accordance with clause 17a of B.S. 449.

In pairs of angles, tees or channels arranged in contact back to back or separated by a distance not exceeding the aggregate thickness of the connected parts, tacking bolts or welding, with solid spacing pieces where the parts are separated, should be provided along the members at intervals not exceeding 1 m. Where the gap between pairs of angles or tees is excessive, presumably greater than that indicated in the previous paragraph, the simplified rules given above are not appropriate. No guidance is given in this respect, and it is suggested that the design follows the principles outlined for channels.

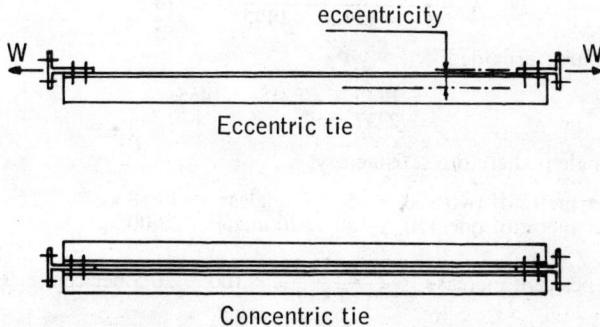

Eccentric tie

Concentric tie

Fig. 31

Figure 31 shows typical concentric and eccentrically arranged tie members.

A tension member carries an axial load of 325 kN. Design this tie using grade 43 steel, and assume that it consists of a pair of angles symmetrical about a 10 mm thick gusset plate, short leg connected by 20 mm diameter bolts.

SOLUTION

From Table 19 (B.S. 449), $p_t = 155$ N/mm^2

Net area required, $A_n = \dfrac{325 \times 10^3}{155} = 2096 \cdot 7$ mm^2

Try two $100 \times 65 \times 8$ angles—gross area $= 2530$ mm^2. Deduct one hole from each angle, and actual net area provided is

$$A_n = 2530 - (2 \times 22 \times 7 \cdot 8) = 2178 \text{ mm}^2$$

Selected angles are satisfactory. Provide solid spacers 75 mm square and 10 mm thick at not more than 1 m intervals, and weld to angles.

EXAMPLE 18

Redesign the tie in the previous example, assuming a single equal angle and using grade 43 steel. Compare masses.

SOLUTION

Table 19 applies, net area required $A_n = 2096 \cdot 7$ mm^2 (as before).

Try one $150 \times 150 \times 10$ angle—gross area $= 2930$ mm^2, therefore gross area of each leg is clearly 1465 mm^2. Net area of connected leg is

$$A_1 = 1465 - (22 \times 10) = 1465 \text{ mm}^2$$

Reduction factor

$$k = \frac{3 \times 1245}{(3 \times 1245) + 1465} = 0 \cdot 718$$

Actual net area provided

$$A'_n = 1245 + (0 \cdot 718 \times 1465)$$
$$= 2297 \text{ mm}^2$$

Selected angle is therefore satisfactory.

Mass per metre of two $100 \times 65 \times 8$ angles $= 19 \cdot 88$ kg
Mass per metre of one $150 \times 150 \times 10$ angle $= 23 \cdot 00$ kg

$$\text{per cent increase} = \frac{23 - 19 \cdot 88}{19 \cdot 88} \times 100 = 20 \cdot 5 \text{ per cent.}$$

This tie is not only less economical in terms of main material consumption, but will require a larger number of bolts to form the connections.

DESIGN OF COMPRESSION MEMBERS

Compression members may consist of universal columns, universal beams, joist sections, channels or angles. The problem surrounding the design of columns is essentially one of stability. Much research and investigation has been carried out in the past and still continues.

In this part discussion will centre round rolled profiles of the type first mentioned. Plate columns having a cross-section similar to a beam will be covered by the same rules discussed later. Concrete-encased columns will be dealt with in the chapter on composite design. Box-type sections are specifically excluded as no firm rules exist in the current B.S. 449.

Basic notions of column behaviour

Euler is credited with the first mathematically derived column formula. It was assumed that initially the column was perfectly straight and the load absolutely concentric. He deduced that when the applied load reached a certain value the column would bend out of straightness into a sine wave and provided the load did not exceed

$$P_{cr} = \frac{\pi^2 EI}{L^2}$$

the column would remain stable and in equilibrium in its deformed shape. This formula presupposes indefinite elasticity and ignores the existence of yield stress, and is therefore valid only for slender columns. Recognising the limitations various authorities have developed expressions to represent practical column behaviour. Such a practical formula is that due to both Perry and Robertson and is named the Perry–Robertson formula; it forms the basis of all column design covered by B.S. 449. It accepts the Euler expression and proceeds to cater for certain practical imperfections which are:

1. the steel will have a yield stress;
2. the axial load may not be perfectly concentric;
3. the column may not be absolutely straight.

Other imperfections exist. Truly theoretical column formulae usually defines precise terminal conditions, such as perfect fixity or pin ends, etc., none of which are quite realistic in practice. Experiment has shown that rolled and welded sections contain residual stresses which influence behaviour.

For the benefit of the reader the above comments serve to highlight the problem of column design and introduce what follows. Many textbooks on advanced strength of materials deal with columns in great detail and should be consulted for a deeper understanding of their behaviour.

Practical column design

A column may be a single component or two or more components arranged parallel to each other and suitably connected together. Rules for design are given in B.S. 449 and are largely empirical but give satisfactory results. Load patterns for column design are defined as follows:

1. axially loaded—the load is assumed to act through the geometric axis of the column profile;
2. eccentrically loaded—the load is assumed by circumstance to be displaced some distance from the geometric axis of the column profile;
3. moment loading—an applied moment due to frame action (rigid frameworks) distinct from that due to eccentricity of load.

It should be noted that in definition 1 the empirical location of the axial load is catered for in the imperfection allowance built in to the column formula. No assumption need therefore be made other than it is said to be axial, otherwise definition 2 will apply.

Slenderness ratios and effective lengths

If in Euler's formula P_{cr} is replaced by Af_{cr} and I by Ar^2 the expression becomes

$$f_{cr} = \frac{\pi^2 E}{(qL/r)^2}$$

Thus it will be seen that since $\pi^2 E$ is constant the critical stress f_{cr} varies inversely with the quantity $(qL/r)^2$. This is important because unlike ties the stress now depends on the length of the member and the factors q and r. The quantity qL/r is defined as the effective slenderness ratio of the column. Factor q takes into account the degree of terminal (end) restraint given to the column, thus qL is defined as the effective length and is given the single symbol l. Hence

$$\text{slenderness ratio} = \frac{\text{effective length}}{\text{radius of gyration}} = \frac{l}{r}$$

The effective length factor q in terms of the actual column length L measured centre to centre of intersections with supporting members for both theoretical and practical (design) conditions is given in the table below.

Case	Degree of end restraint	Theoretical	Practical
1	Effectively held in position and restrained in direction at both ends	0·5L	0·7L
2	Effectively held in position at both ends and restrained in direction at one end	0·71L	0·85L
3	Effectively held in position at both ends, but not restrained in direction	1·0L	1·0L
4	Effectively held in position and restrained in direction at one end, the other end partially restrained in direction but not held in position	—	1·5L
5	Effectively held in position and restrained in direction at one end, the other end completely free	2·0L	2·0L

Permissible axial compressive stress

A slightly modified Perry–Robertson formula is used to derive the permissible axial compressive stress and takes the form

$$K_2 p_c = \frac{f_y + (\eta + 1)f_{cr}}{2} - \sqrt{\left[\left(\frac{f_y + (\eta + 1)f_{cr}}{2}\right)^2 - f_y f_{cr}\right]},$$

where
K_2 = load factor, taken at 1·7;
f_{cr} = Euler stress as previously discussed;
f_y = yield stress of the material grade used;
$\eta = 0·3(l/100r)^2$.

The factor η is the imperfection factor which takes into account lack of straightness and want of perfect concentricity of the applied load. Values of p_c are given in Table 17 of B.S. 449 for three grades of steel. This formula appears to be formidable, but in fact represents the solution of a quadration equation derived from $f = P/A + M/Z$. Since the column is initially bent this is magnified by the load P and produces the moment M, thus both axial and bending stresses are produced. When the sum of these stresses on the concave side of the column reaches $f = f_y$ the limit of elastic usefulness is attained.

Eccentricities

In simple design the reactions from incoming beams are assumed to be eccentric to the column axis. The position of the reaction from a beam should be taken as its centre of bearing or 100 mm in front of the column flange or web receiving this beam, whichever gives the worst effect.

For beams supported on column cap plates the reaction should be applied at the face of the column towards the direction of the span. Should the plan dimensions of the cap plate exceed that of the column and this plate is stiff or is stiffened, the reaction will be towards the edge of this plate.

Roof trusses bearing on column caps are assumed to offer no eccentricity. It is to be taken that these bearings are simple and that the connections develop no moment.

Eccentricity moments

Eccentric reactions introduce bending moments into the column. In cap-type connections the whole of the moment is resisted by the column and takes the value

$$M = Re$$

In continuous columns the reaction moment from an incoming beam is shared at the level of that beam between the upper and lower lengths of the column in proportion to their stiffness. The shared moment is not assumed to be carried over to other levels.

The unshared reaction moment is

$$M = Re$$

If the stiffness of the upper and lower column lengths at the level considered are K_u and K_L and M_u and M_L represent the shared moment values then

$$M_u = Re\frac{K_u}{K_u + K_L}$$

and

$$M_L = Re\frac{K_L}{K_u + K_L}$$

when K_u/K_L or $K_L/K_u < 1.5$ it may be assumed that

$$M_u = M_L = 0.5Re.$$

Actual and permissible stress—ratios

To design a column pre-knowledge of its size is required. Without experience this is a shot in the dark, but after some practice it becomes an intelligent guess.

For all columns the actual compressive stresses is merely

$$f_c = \frac{P}{A},$$

and the permissible compressive stress p_c comes from Table 17 of B.S. 449, taken in relation to the slenderness ratio measured about the weakest axis of the column. If there are no eccentricities or applied moments then

$$\frac{f_c}{p_c} \leqslant 1$$

The actual compressive bending stress f_{bc} due to an applied or eccentricity moment is found as in beam design, but the permissible compressive bending stress p_{bc} will depend upon l/r and D/T of the column.

Under the combined effect of axial load and bending moment the stresses should be proportioned in the form

$$\frac{f_c}{p_c} + \frac{f_{bc}}{p_{bc}} \leqslant 1$$

Should there be simultaneous bending about two principal column axes the values of p_{bc} for each axis may not be equal then

$$\frac{f_c}{p_c} + \left(\frac{f_{bc}}{p_{bc}}\right)_{xx} + \left(\frac{f_{bc}}{p_{bc}}\right)_{yy} \leqslant 1$$

Special restrictions

Columns designed from a single universal beam, column or channel suffer no special restrictions other than an upper limit on the slenderness ratio which must not exceed 180.

If a column consists of two of the above members and is regarded as one for design additional rules must be complied with to ensure suitability. In general both components should be of the same type and size and must be connected together at their remote ends and at intermediate positions. Properly designed welded battens are used for this purpose and spaced so that each component of the column will carry one half of the total load. Detailed conditions are given in clauses 36 and 37 of B.S. 449.

Angles used as discontinuous compression members are also subject to special provisions.

A single angle connected at both ends by only one bolt is assigned an effective length equal to its actual length, and the calculated axial stress f_c should not exceed 80 per cent of the Table 17 value. Moreover, the l/r value should in no case exceed 180.

Doubling the number of bolts at each end or providing the equivalent in weld metal allows the effective length to be reduced to $0.85L$, with full advantage being taken of permissible stress.

Twin angles having at least two bolts in each end have an effective length of $0.85L$, but must be connected together along their length if the pair are treated as a single member. The spacing of the connectors has the same significance as that explained earlier and should comply with clause 37. Under the provisions of this clause it is assumed that the members will either be in close contact or separated by a small distance (say the thickness of a gusset plate). If this distance is large it is suggested that the battened compression member rules in clause 36 will be appropriate.

Construction details

Where compression members require bases these should be stiff enough to spread the load through the foundation without causing overstress. Bases may be either of the slab type or built up from plates. When the pressure on the underside of the base is uniform the slab type may prove to be more economical. The thickness t of the slab should not be less than

$$t = \sqrt{\left[\frac{3w}{p_{bct}} \left(A^2 - \frac{B^2}{4} \right) \right]},$$

where A and B are the greater and lesser projections of the base beyond the column; w is the calculated pressure; and p_{bct} the design stress which is limited to 185 N/mm² for all grades of steel. At least two 16 mm diameter holding-down bolts should be provided, together with a grouting hole.

Built-up bases are to be preferred when both axial load and bending moment are present. Two cases are to be considered:

1. the combined effect of the axial load and moment produce a uniformly varying pressure (compression) over the entire underside of the base;
2. the combined effect, as above, produces a zone of compression and a zone of tension.

It is better to use four bolts for this type of base, but where tension is developed the bolts in this zone should be checked and the size and number increased if necessary.

Because of the combined effect the base size must be assumed. The maximum and minimum stresses may be found from

$$f = \frac{P}{A} \pm \frac{M}{Z}$$

The maximum value should not exceed that for the concrete forming the foundation. If f is always compressive (1) above, applies. If f becomes tensile, *i.e.* $P/A - M/Z$ is negative, (2) applies and this expression is not relevant. From Fig. 32 proceed as follows: (C = total compression in shaded area; T = total tensile to be resisted by holding-down bolts; b = width of base).

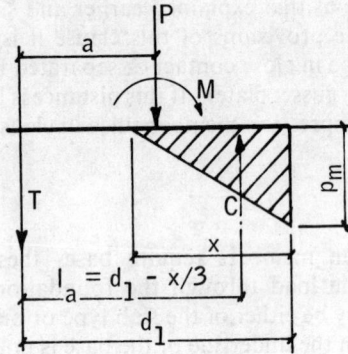

FIG. 32

For equilibrium $\Sigma V = 0$ then

$$T + P - C = 0, \quad . \quad . \quad . \quad . \quad . \quad (1)$$

but

$$C = p_m \cdot \frac{bx}{2},$$

combine with eqn. (1) to give

$$T + P - p_m \cdot \frac{bx}{2} = 0$$

But since $\sum M = 0$, then, taking moments about T,

$$M + Pa - Cl_a = 0 \quad . \quad . \quad . \quad . \quad (2)$$

substitute for C and l_a and eqn. (2) becomes

$$M + Pa - p_m \cdot \frac{bx}{2}\left(d_1 - \frac{x}{3}\right) = 0$$

Assign a value to p_m consistent with the grade of foundation concrete and solve eqn. (2) for x. Substitute for x in eqn. (1) and find T.

A final general comment. It is important to remember that the design of a column and its base are interconnected. Whatever assumption is made relating to the column effective length the base must be designed with this in mind.

EXAMPLE 19

A simple strut resists an axial force of 18·8 kN. Its actual length, centre to centre of end connections, is 2·0 m. Using grade 43 steel determine the size of a single equal angle assuming

(a) single bolting each end;
(b) double bolting each end.
Compare their masses.

SOLUTION

(a) Try 60 × 60 × 8 angle ($A = 903$ mm² $r_{min} = 11·6$ mm). Actual stress

$$f_c = \frac{18·8 \times 10^3}{903} = 20·8 \text{ N/mm}^2$$

The slenderness ratio must be calculated on the actual length (see clause 30c(i)) then

$$\frac{l}{r} = \frac{2000}{16·6} = 172·4$$

From Table 17, $p_c = 31$ N/mm². Now f_c must not exceed $0·8p_c$ also by the above clause, then

$$\frac{f_c}{p_c} = \frac{20·8}{31} = 0·67 < 0·8$$

Angle section selected is therefore satisfactory.
(b) Try 60 × 60 × 5 angle ($A = 582$ mm² $r_{min} = 11·7$ mm). Actual stress is

$$f_c = \frac{18·8 \times 10^3}{582} = 32·3 \text{ N/mm}^2$$

The slenderness ratio may be calculated on $0·85L$ per clause 30c(i)

$$\frac{l}{r} = \frac{0·85 \times 2000}{11·7} = 145·3$$

From Table 17, $p_c = 43$ N/mm², and f_c must not exceed p_c or

$$\frac{f_c}{p_c} = \frac{32·3}{43} = 0·75 < 1$$

Angle section selected is satisfactory.
Comparing masses per unit length:

(a) One 60 × 60 × 8 angle—mass = 7·09 kg.
(b) One 60 × 60 × 5 angle—mass = 4·57 kg.

Therefore by using at least two bolts in each end connection saving in material is

$$\frac{7 \cdot 09 - 4 \cdot 57}{4 \cdot 57} \times 100 = 55\% \text{ in terms of the lighter,}$$

or

$$\frac{7 \cdot 09 - 4 \cdot 57}{7 \cdot 09} \times 100 = 35 + \text{ in terms of the heavier.}$$

EXAMPLE 20

A compression member supports an axial load of 102 kN, and the actual length between centres of end connections is 3·5 m. Design a pair of angles using grade 50 steel for:

(a) short legs connected either side a 10 mm gusset plate;
(b) long legs connected either side a 10 mm gusset plate.

Design the battening for case (a) and compare unit masses

SOLUTION

(a) Try two 100 × 65 × 10 angles (area 3120 mm). From section tables $r_{xx} = 18 \cdot 1$ mm, $r_{yy} = 49 \cdot 8$ mm.

Actual stress, $\qquad f_c = \dfrac{102 \times 10^3}{3120} = 32 \cdot 7 \text{ N/mm}^2$

Clause 30c(ii) permits an effective length between 0·7 and 0·85. Using the latter

$$\frac{l}{r_{xx}} = \frac{0 \cdot 85 \times 3500}{18 \cdot 1} = 164,$$

From Table 17b $p_c = 37 \text{ N/mm}^2$, giving

$$\frac{f_c}{p_c} = \frac{32 \cdot 7}{37} = 0 \cdot 88 < 1$$

Angles selected are just satisfactory.

(b) Try two 80 × 60 × 6 angles (area = 1620 mm). From section tables $r_{xx} = 25 \cdot 2$ mm, $r_{yy} = 25 \cdot 7$ mm.

Actual stress, $\qquad f_c = \dfrac{102 \times 10^3}{1620} = 62 \cdot 9 \text{ N/mm}^2,$

$$\frac{l}{r_{xx}} = \frac{0 \cdot 85 \times 3500}{25 \cdot 2} = 118,$$

from Table 17b

$$p_c = 69 \text{ N/mm}^2,$$

giving

$$\frac{f_c}{p_c} = \frac{62 \cdot 9}{69} = 0 \cdot 91 < 1$$

which is just satisfactory.

Battening for case (a) design

Clause 37 applies—the slenderness ratio of each angle measured between centres of connector packs (battening) must not exceed 40 or 0·6 of the most unfavourable l/r for the whole strut, then

$$0·6\frac{l}{r_{xx}} = \frac{0·6 \times 0·85 \times 3500}{18·1} = 98·6,$$

or 40.

Using 40 the maximum distance between packs is related to the least radius of gyration of one $100 \times 65 \times 10$ angle, *i.e.* $r_{vv} = 13·9$ mm, then

$$l = 13·9 \times 40 = 556 \text{ mm}$$

The actual distance will be somewhat less than this depending upon end connections and will be resolved at the detailing stage. Use 60 mm square packs 10 mm thick inserted between angles and welded to them.

Comparing masses—per unit length:

(a) Two $100 \times 65 \times 10$ angles—mass = 24·6 kg.
(b) Two $80 \times 60 \times 6$ angles—mass = 12·7 kg.

Clearly the advantage lies with longer legs connected, *i.e.* case (b) design.

EXAMPLE 21

The internal row of columns in a two-storey warehouse extend between the ground and first floor only. The storey height is 5·0 m. Construction requirements show that the column finishes 300 mm below the ground-floor slab and is connected to a slab-type base plate, while the top of the column is flush with the underside of the first-floor slab which is 200 mm thick. Loading and incoming beam sizes are as follows:

To column flange (north) $914 \times 305 \times 224$ U.B. 410 kN
To column flange (south) $686 \times 254 \times 152$ U.B. 270 kN
To column web (east) $381 \times 152 \times 52$ U.B. 69 kN
To column web (west) $533 \times 163 \times 73$ U.B. 141 kN

Design a grade 43 column and base plate.

SOLUTION

For strength calculations take length of column between base and neutral axis of east beam so that

$$L = 5000 + 300 - 200 - 190 = 4910 \text{ mm}$$

Since the beams connecting to the column web faces are not approximately equal in size or reaction the effective length factor should be taken as unity.

Try $305 \times 305 \times 97$ U.C. $A = 12\,330$ mm², $r_{yy} = 76·8$ mm, $D/T = 20$, $Z_{xx} = 1442 \times 10^3$ mm³, $Z_{yy} = 476·9 \times 10^3$ mm³.

$$\frac{l}{r_{yy}} = \frac{4910}{76·8} = 64$$

giving p_c from Table 17a of 122 N/mm², and p_{bc} from Table 3 of 165 N/mm²

Loads to be carried:

From north flange	410
From south flange	270
From east web	69
From west web	141
Add sw + encasement	8
Total	898 kN

Actual stress

$$f_c = \frac{898 \times 10^3}{12\,330} = 72 \cdot 9 \text{ N/mm}^2,$$

giving

$$\frac{f_c}{p_c} = \frac{72 \cdot 9}{122} = 0 \cdot 596 < 1$$

Stresses due to bending

1. About xx-axis:

Out-of-balance load $= 410 - 270 = 140$ kN

Eccentricity $e = 100 + \dfrac{307 \cdot 8}{2} = 254$ mm

$$M_{xx} = 140 \times 254 \times 10^3 = 35 \cdot 56 \times 10^6 \text{ Nmm}$$

$$f_{bc,xx} = \frac{35 \cdot 56 \times 10^6}{1442 \times 10^3} = 24 \cdot 7 \text{ N/mm}^2$$

2. About yy-axis:

Out-of-balance load $= 141 - 69 = 72$ kN

Eccentricity $e = 100 + \dfrac{9 \cdot 9}{2} = 105$ mm

$$M_{yy} = 72 \times 105 \times 10^3 = 7 \cdot 56 \times 10^6 \text{ Nmm}$$

$$f_{bc,yy} = \frac{7 \cdot 56 \times 10^6}{476 \cdot 9 \times 10^3} = 15 \cdot 9 \text{ N/mm}^2$$

Now since permissible stress about both axes is the same, actual stresses due to bending may be summed and ratioed thus

$$f_{bc,xx} + f_{bc,yy} = 24 \cdot 7 + 15 \cdot 9 = 40 \cdot 6 \text{ N/mm}^2$$

and

$$\frac{f_{bc}}{p_{bc}} = \frac{40 \cdot 6}{165} = 0 \cdot 246 < 1,$$

whence

$$\frac{f_c}{p_c} + \frac{f_{bc}}{p_{bc}} = 0 \cdot 596 + 0 \cdot 246 = 0 \cdot 842 < 1$$

Column size selected is satisfactory.

Base plate

Taking safe bearing pressure as 5 N/mm², least area required is

$$\frac{898 \times 10^3}{5} = 179{\cdot}6 \times 10^3 \text{ mm}^2$$

Try a square base 425 mm. Greater projection,

$$A = 0{\cdot}5(425 - 304{\cdot}8) = 60{\cdot}1 \text{ mm},$$

lesser, projection,

$$B = 0{\cdot}5(425 - 307{\cdot}8) = 58{\cdot}6 \text{ mm}$$

Actual base pressure,

$$w = \frac{898 \times 10^3}{425^2} = 4{\cdot}97 \text{ N/mm}^2$$

and

$$p_{\text{bc t}} = 185 \text{ N/mm}^2 \text{ from Table 2, then}$$

$$t = \sqrt{\left[\frac{3 \times 4{\cdot}97}{185}\left(60{\cdot}1^2 - \frac{58{\cdot}6^2}{4}\right)\right]}$$

$$= 14{\cdot}89 \text{ (use 15 mm)}$$

Note: End of column to be machined flat. Slab to be machined flat if surface is untrue. Provide nominal fillet weld all round column for location only.

EXAMPLE 22

Two plate girders each 1·5 m deep and 500 mm wide, producing reactions of 2000 kN respectively are to be connected to the major axis of a column whose actual height is 14 m. Both minor axis beams are 459 × 152 × 52 U.B. with reactions of 165 kN each. Design a suitable column in grade 50 steel having a built-up base.

SOLUTION

Total load carried, excluding self-mass is 4330 kN. A single 356 × 406 × 340 U.C. would suffice, however, as the connection to the plate girders will be difficult examine a compound column using two U.B.s. For compound columns it is better to arrange for the moments of inertia about both principal axes to be equal or nearly so. This gives

$$2I_{xx} = 2[I_{yy} + Ad^2]$$

Try two 457 × 191 × 98 U.B.s, $I_{xx} = 45\,653 \times 10^4$ mm⁴, $I_{yy} = 2216 \times 10^4$ mm⁴; $A = 125{\cdot}2 \times 10^2$ mm², $r_{xx} = 191{\cdot}0$ mm, $r_{yy} = 42{\cdot}1$ mm.

Solving for d in the above expression gives

$$d = 10\sqrt{\left(\frac{45\,653 - 2216}{125{\cdot}2}\right)} = 186 \text{ mm},$$

so that the spacing centre to centre of U.B.s is 372 mm, and of course

$$r_{yy} = r_{xx} = 191 \text{ mm}$$

For design the net column length is 14 m less half the depth of the incoming beams, that is, major axis

$$L = 14 - 0.75 = 13.25 \text{ m},$$

minor axis

$$L = 14 - 0.23 = 13.77 \text{ m}$$

The effective length factor may be taken as 0·7 on account of the stiff built-up base and the balanced nature of the top beams. Then

$$\frac{l}{r_{xx}} = \frac{0.7 \times 13.25 \times 10^3}{191} = 48.5,$$

$$\frac{l}{r_{yy}} = \frac{0.7 \times 13.77 \times 10^3}{191} = 50.4$$

From Table 17b, $p_c = 183.6 \text{ N/mm}^2$. Actual stress f_c after allowing 30 kN for self-weight is

$$f_c = \frac{4360 \times 10^3}{2 \times 125.2 \times 10^2} = 174.2 \text{ N/mm}^2$$

which is satisfactory.

This column will require battens complying with clause 36. Batten spacing —since $l/r_{yy} > 0.8 l/r_{xx}$ the l/r value of each U.B. between battens must not exceed 40 or $0.6 \, l/r_{yy}$. Now $0.6 \times 50.4 = 30.24 < 40$ (use 30·24) then

$$\frac{l}{r_{yy}} = 30.24$$

giving

$$l = 30.24 \times 42.1 = 1272 \text{ mm}$$

Make top battens 2·0 m deep to receive plate girders plus allowance for built-up brackets. Use base cheek plates as bottom battens and make these 0·5 m deep. Then net length for other battens is

$$L = 14 - (2 + 0.5) = 11.5 \text{ m}$$

The effective length of each intermediate batten (clause 36e) must not be less than the larger of

$$l_b = 0.75 \times 372 = 279 \text{ mm},$$

or

$$l_b = 2.0 \times 191 = 382 \text{ mm}$$

Assuming seven intermediate battens each 380 mm long their spacing will be

$$\frac{11\,500 - (7 \times 380)}{8} = 1105 \text{ mm} < 1272$$

Provide battens in pairs 380 mm long spaced at 1105 mm longitudinally. Clause 36c requires that each pair of battens must resist a bending moment and shear force due to a transverse shear of $2\frac{1}{2}$ per cent of the total axial load in the column; therefore

$$F_q = \frac{4360}{40} = 109 \text{ kN}$$

from clause 30d the longitudinal shear $F_1 = F_q \cdot d/na$ and the bending moment $M = F_q d/2n$ where (*see* Fig. 33)

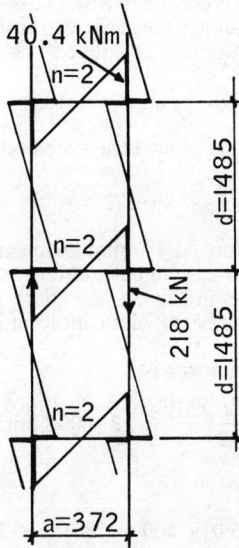

moments and shears
in batten plates

FIG. 33

d = longitudinal distance centre to centre of battens—1105 + 380
    = 1485 mm
a = transverse distance between weld groups = 372 mm
n = number of parallel planes of battens = 2

then

$$F_1 = \frac{109 \times 1486}{2 \times 372} = 218 \text{ kN}$$

and

$$M = \frac{109 \times 1 \cdot 485}{2 \times 2} = 40 \cdot 4 \text{ kNm}$$

Taking $p_q = 140 \text{ N/mm}^2$ and $p_{bc} = 230 \text{ N/mm}^2$ the minimum thickness of batten plate is

$$140 \times 380 \times t = 218 \times 10^3$$

giving

$$t = 4 \cdot 1 \text{ mm}$$

or

$$\frac{230 \times 380^2 \times t}{6} = 40\cdot4 \times 10^6$$

giving

$$t = 7\cdot3 \text{ mm},$$

use 10 mm thick battens (this also complies with clause 30g) with fully continuous profile fillet welds to column. These welds must be designed for the longitudinal shear and bending moment calculated above.

Built-up base

Using a bearing pressure of 5000 kN/m² least area required is

$$A = \frac{4360}{5000} = 0\cdot872 \text{ m}^2$$

Make the base rectangular 1000 mm × 880 mm giving an actual base pressure of 4·96 N/mm², say 5·0 N/mm² for ease of calculation.

Provide base stiffeners as shown in Fig. 34 and check base thickness along longer edge by treating as a simple beam with cantilever ends. Assume beam has unit width.

At the stiffeners the bending moment is

$$M = \frac{5\cdot0 \times 220^2}{2} = 121\ 000 \text{ Nmm}$$

and at the centre of the long edge

$$M = \frac{5\cdot0 \times 500^2}{2} - (5\cdot0 \times 500 \times 280) = -75\ 000 \text{ Nmm}$$

Taking $p_{bc} = 230$ N/mm², then

$$121\ 000 = 230 \times \frac{t^2}{6}$$

from which $t = 56\cdot2$ mm, say 60 mm.

Check strength of combined cheek plate and base as an inverted tee beam along plane XX. Load carried on one cheek plate unit is

$$5\cdot0 \times 440 \times 218 \times 10^{-3} = 479\cdot6 \text{ kN},$$

$$\therefore \quad M = 479\cdot6 \times \frac{218 \times 10^{-3}}{2} = 52\cdot3 \text{ kNm}.$$

Making thickness of cheek plate 15 mm (clause 28b(iii) by inference) the calculated properties of this inverted tee beam are: $\bar{y} = 92$ mm, $I_{xx} = 622 \times 10^6$ mm⁴, $Z_1 = 1329 \times 10^3$ mm³, $Z_2 = 6765 \times 10^3$ mm³; then

$$f_q = \frac{479\cdot6 \times 10^3}{480 \times 15} = 66\cdot5 \text{ N/mm}^2$$

$$f_{bc} = \frac{52\cdot3 \times 10^6}{1329 \times 10^3} = 39\cdot4 \text{ N/mm}^2$$

$$f_{bt} = \frac{52\cdot3 \times 10^6}{6765 \times 10^3} = 7\cdot7 \text{ N/mm}^2$$

which are very low.

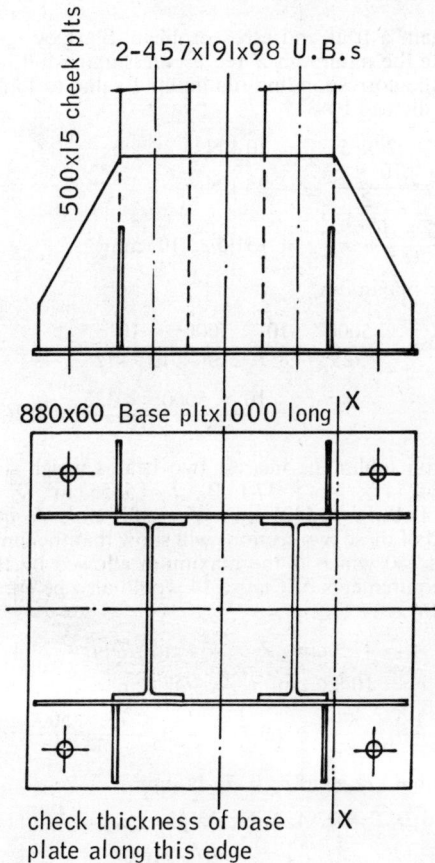

2-457x191x98 U.B.s

500x15 cheek plts

880x60 Base pltx1000 long

check thickness of base
plate along this edge

Built-up base

Fig. 34

The horizontal shear between the base and cheek plate required for welding
is

$$qb = \frac{QA\bar{Y}}{I} = \frac{479 \cdot 6 \times 10^3 \times (440 \times 60) \times 62}{622 \cdot 1 \times 10^6} = 1261 \text{ N/mm}$$

Use 6 mm fillet welds to both sides of cheek plates.

Machine end of column and underside of cheek plates to achieve dead
bearing to base plate.

EXAMPLE 23

A column in a simple factory building is 5 m high and is restrained by
sheeting rails at approximately 1·25 m centres about its minor axis. The load
from the roof, cladding, rails and self weight is 60 kN, and there is a horizontal wind force of 2kN per metre height of column. Design the column and
and its base plate together with holding down bolts. Use Grade 43 steel.

SOLUTION

This is once again a trial and error problem. Start by ignoring the axial force and calculate the minimum Z required assuming full bending stress of 165 N/mm^2 and the corresponding minimum I value to limit the deflection to column height divided by 325.

Total lateral load $= 2 \times 5 \qquad = 10$ kN

Bending moment $= \dfrac{10 \times 5}{2} \qquad = 25$ kNm

Minimum $Z = \dfrac{25 \times 10^6}{165} \qquad = 151 \cdot 5 \times 10^3$ mm^3

For minimum I_{xx}, maximum

permitted deflection $= \dfrac{5000}{325} = \dfrac{10 \times 5000^3 \times 10^3}{8 \times 210 \times 10^3 \times I}$

Minimum $I_{xx} \qquad = \dfrac{10 \times 5000^2 \times 325}{8 \times 210} = 4836 \cdot 3 \times 10^4$ mm^4

From the section tables the nearest two beams which satisfy both these requirements are $254 \times 146 \times 37$ U.B., $I = 5556$ cm^4, $Z = 434$ cm^3, and $305 \times 102 \times 28$ U.B., $I = 5421$ cm^4, $Z = 351$ cm^3. A quick calculation against the second of these two sections will show that the limiting slenderness ratio just exceeds 180 which is the maximum allowed by B.S. 449; further the stress ratio requirements of Clause 14 would also be exceeded.
Try the $254 \times 146 \times 37$ U.B.,

$$A = 47 \cdot 5 \text{ cm}^2, \ Z = 434 \text{ cm}^3, \ D/T = 23 \cdot 4$$

$$r_x = 10 \cdot 82 \text{ cm}, \ r_y = 3 \cdot 47 \text{ cm}$$

$$\frac{l}{r_x} = \frac{1 \cdot 5 \times 5000}{108 \cdot 2} = 69 \cdot 3, \qquad \frac{l}{r_y} = \frac{0 \cdot 75 \times 5000}{34 \cdot 7} = 108$$

For $l/r = 108$, Table 17a gives $p_c = 71$ N/mm^2

For $l/r = 108$ and $D/T = 23 \cdot 4$, Table 3a gives $p_{bc} = 149$ N/mm^2

direct stress $f_c \qquad = \dfrac{60 \times 10^3}{4750} = 12 \cdot 63$ N/mm^2

bending stress $f_{bc} \qquad = \dfrac{25 \times 10^6}{434 \times 10^3} = 57 \cdot 6$ N/mm^2

so that $\qquad \dfrac{f_c}{p_c} + \dfrac{f_{bc}}{p_{bc}} = \dfrac{12 \cdot 63}{71} + \dfrac{57 \cdot 6}{149}$

$$= 0 \cdot 18 + 0 \cdot 39 = 0 \cdot 56 < 1 \cdot 25$$

This ratio appears very small and uneconomic, but deflection controls the design, not strength.

Design of baseplate

From the foregoing it will be seen that the column is in tension to the sum of $57 \cdot 6 - 12 \cdot 6 = 45$ N/mm^2, and this serves as an immediate indication that the base will need tension bolts for holding down.
Try a base plate 450 mm long by 200 mm wide with four holding-down bolts. Details are given in Fig. 35.

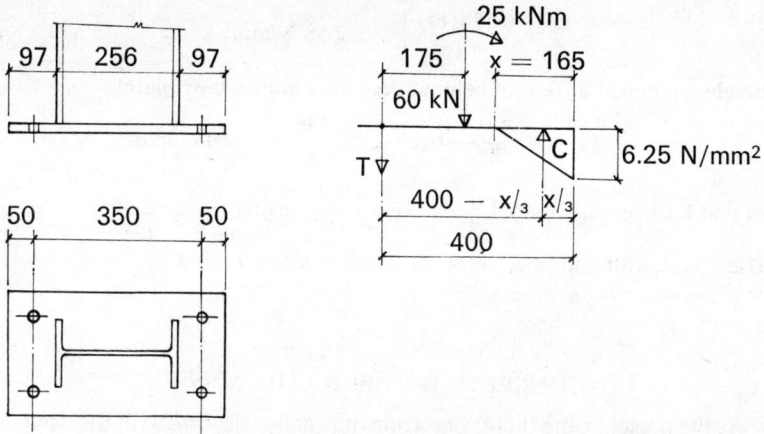

Fig. 35

Taking the concrete pressure at 5 N/mm², plus 25 per cent for wind

$$C = 6 \cdot 25 \times 200 \times \frac{x}{2} = 625x \text{ N}$$

then
$$T + 60\ 000 - 625x = 0 \tag{1}$$

also
$$25 \times 10^6 + 60 \times 175 \times 10^3 - 625x \left(400 - \frac{x}{3}\right) = 0 \tag{2}$$

From (2), $35 \cdot 5 \times 10^6 - 0 \cdot 25x \times 10^6 - 208 \cdot 3x^2 = 0$

By quadratics this solves to give $x = 165$ mm

So that in (1) $T + 60\ 000 - 625 \times 165 = 0$,
giving T = 43 125 N

Permissible stress in Grade 4·6 bolts plus wind = 110 + 25%
= 137·5 N/mm tension, and 80 + 25% = 100 N/mm shear.
Tension is resisted by two bolts,
  Try M20 black bolts, $A_s = 245$ mm² $A_g = 314$ mm².

$$\text{Tensile stress } f_t = \frac{43\ 125}{2 \times 245} = 88 \cdot 0 \text{ N/mm}^2$$

The total lateral wind force of 10 kN is resisted by all four bolts as a shear action.

$$\text{shear stress } f_s = \frac{10 \times 10^3}{4 \times 314} = 7 \cdot 96 \text{ N/mm}^2$$

$$\therefore \frac{f_t}{p_t} + \frac{f_s}{p_s} = \frac{88 \cdot 0}{137 \cdot 5} + \frac{7 \cdot 96}{100} = 0 \cdot 72 < 1 \cdot 4, \text{ satisfactory.}$$

To calculate base plate thickness use $p_{bc} = 165$ N/mm² plus 25 per cent for wind pressure at face of column

$$= \frac{6 \cdot 25 \,(165-97)}{165} = 2 \cdot 58 \text{ N/mm}^2$$

Bending moment at face of column per unit width of base plate

$$= 2 \cdot 58 \times \frac{97^2}{2} + (6 \cdot 25 - 2 \cdot 58) \frac{97^2}{3} = 23 \ 648 \text{ Nmm}$$

to find thickness t, required, $\dfrac{t^2}{6} = \dfrac{23 \ 648}{1 \cdot 25 \times 165}$, giving $t = 26 \cdot 2$ mm.

Use a base with a finished thickness of 30 mm.

CONCRETE ENCASED COLUMNS

At the present time there is no code of practice dealing with the design of composite columns. Clause 30b of B.S. 449 does, however, give rules for elastically designed concrete encased columns but the following restrictions apply:

1. Box columns are specifically excluded, as are angles and tees.
2. The overall dimension of a steel column must not exceed 1000 mm $\times$ 500 mm in section.
3. Channels if used must be either back to back in contact, or if separated not less than 20 mm apart nor more than half their depth apart, and properly laced or battened together per clauses 35 and 36.
4. Steel must be unpainted and solidly encased in dense concrete with 10 mm aggregate and having a 28-day strength of at least 21 N/mm².
5. Concrete cover must be at least 50 mm to give a minimum width of solid encasement of $B + 100$ ($B =$ flange width). Any cover exceeding 75 mm must be ignored.
6. Links at least 5 mm diameter and not more than 150 mm pitch, tied to at least four longitudinal bars, must be provided and arranged symmetrically within the cover.
7. Only axial loads may be carried by an encased column and moments must be resisted by the steel section alone.
8. The radius of gyration r_y may be increased to $0 \cdot 2 \,(B + 100)$ but r_x is as for the uncased section.
9. The calculated axial load on the cased column must not exceed twice that for the uncased section, nor must the slenderness ratio for the uncased column exceed 250.
10. The stress on the concrete encasement must not exceed the permissible axial stress in the steel divided by $0 \cdot 19$ times the numerical value of p_{bc} from Table 2.

Hence the safe working load of an encased steel column is the lesser of the two values:

$$W_c = p'_c \left[A_s + \frac{A_c}{0.19p_{bc}} \right]$$

or

$$W_u = 2p_c A_s$$

in which p'_c and p_c are the permissible stresses for the cased and uncased sections respectively. A typical cross-section is shown in Fig. 36.

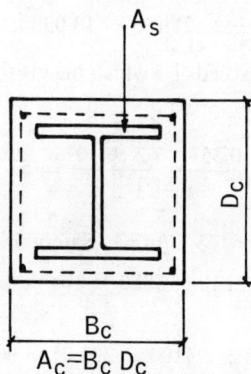

FIG. 36

The combined effect of axial load and bending should be checked to ensure that

$$\frac{P}{W} + \frac{f_{bc}}{p_{bc}} \not> 1$$

EXAMPLE 24

In a six-storey building the external columns receive minor axis reactions of 15 kN from both sides while the major axis reaction is 105 kN. The first-storey column has been designed and yields a $152 \times 152 \times 37$ U.C. cased to 260×260 mm in grade 43.

Design the ground-storey length if the storey height is 3·2 m and all others 3·0 m, assuming a pin base.

SOLUTION

Try a $203 \times 203 \times 46$ U.C., cased to 310×310 mm overall. Direct load to be carried $= 6[105 + 2 \times 15] = 810$ kN. Undistributed reaction moment is

$$M = 105 \times [100 + 101.6] = 21\ 168 \text{ kN mm}$$

This moment must be distributed between the ground- and first-storey column in proportion to their uncased stiffnesses.

$$I_{xx} \text{ of } 152 \times 152 \times 37 \text{ U.C.} = 2218 \times 10^4 \text{ mm}^4,$$
$$I_{xx} \text{ of } 203 \times 203 \times 46 \text{ U.C.} = 4564 \times 10^4 \text{ mm}^4,$$

then

$$K_u = \frac{2218 \times 10^4}{3 \times 10^3} = 7390$$

$$K_{\mathrm{L}} = \frac{4564 \times 10^4}{3 \cdot 2 \times 10^3} = 14\,270$$

$$K_{\mathrm{u}} + K_{\mathrm{L}} = 7390 + 14\,270 = 21\,660$$

whence

$$M_{\mathrm{u}} = \frac{7390}{21\,660} \times 21\,168 = 7238\ \mathrm{kN\,mm}$$

$$M_{\mathrm{L}} = \frac{14\,270}{21\,660} \times 21\,168 = 13\,930\ \mathrm{kN\ mm}$$

Note: The ratio of $K_{\mathrm{L}}/K_{\mathrm{U}}$ exceeds 1·5 which means that the moment cannot be shared equally.

Uncased section

$$\frac{l}{r} = \frac{0 \cdot 85 \times 3 \cdot 2 \times 10^3}{51 \cdot 1} = 53 \cdot 2.$$

From Table 17a
$p_{\mathrm{c}} = 131\ \mathrm{N/mm^2}$ and $p_{\mathrm{bc}} = 165\ \mathrm{N/mm^2}$. Safe working load is

$$W_{\mathrm{u}} = 2 \times 131 \times 5880 \times 10^{-3} = 1540\ \mathrm{kN},$$

cased section,

$$r = 0 \cdot 2 \times [100 + 203 \cdot 2] = 60 \cdot 6\ \mathrm{mm}$$

$$\frac{l}{r} = \frac{0 \cdot 85 \times 3 \cdot 2 \times 10^3}{60 \cdot 6} = 43 \cdot 9$$

From Table 17a $p_{\mathrm{c}} = 137\ \mathrm{N/mm^2}$. Area of cased section,

$$A_{\mathrm{c}} = 310 \times 310 = 96\,100\ \mathrm{mm^2}$$

Equivalent cased area

$$A_{\mathrm{e}} = \frac{96\,100}{0 \cdot 19 \times 165} = 3070\ \mathrm{mm^2}.$$

Total area of cased section:

$$A = 5880 + 3070 = 8950\ \mathrm{mm^2},$$

therefore safe working load is

$$W_{\mathrm{c}} = 137 \times 8950 \times 10^{-3} = 1226\ \mathrm{kN}$$

Since W_{c} is less than W_{u} the cased safe load must be used.
The bending stresses are calculated on the uncased section:

$$f_{\mathrm{bc}} = \frac{13\,930 \times 10^3}{449 \cdot 2 \times 10^3} = 31\ \mathrm{N/mm^2}$$

then

$$\frac{P}{W_{\mathrm{c}}} + \frac{f_{\mathrm{bc}}}{p_{\mathrm{bc}}} \not> 1,$$

and

$$\frac{810}{1226} + \frac{31}{165} = 0 \cdot 662 + 0 \cdot 188 = 0 \cdot 850 < 1.$$

Selected section is satisfactory.

CHAPTER 4

TRUSSES AND BRACING

MANY industrial buildings such as factories and warehouses employ trusswork and bracing in some form. Trusswork is a convenient means of supporting a roof, although this practice has given way to a large extent to portal frame construction. Bracing nevertheless is still required to transfer horizontal forces to the ground, or to provide stability.

METHODS OF ANALYSIS

It is not intended that a lengthy discussion should take place here on methods of analysis, since these are covered in full in elementary textbooks on the theory of structures (*See* Materials and Structures and Theory of Structures in this series). However, after the geometry of a truss has been established, it is normal to identify the load spaces and the member spaces using Bow's notation. After calculating the magnitude of the forces to be carried by the truss and placing these forces in the correct position for analysis, the member forces may be found by using one of the following techniques:

1. graphics *i.e.* force diagrams;
2. method of joint resolution;
3. method of sections;
4. tension coefficients.

The choice of technique is the designer's.

Having found the member forces it will be seen that they divide into two types, *i.e.* tensile and compressive. The design of simple ties and struts was dealt with in the previous chapter—without regard to many other factors which will be discussed in the next section.

DESIGN CONSIDERATIONS

Figure 37 shows a roof truss suitable for a span of about 15—20 m. For the purposes of analysis this truss consists of 27 separate members jointed at their ends (the node points) by frictionless pins. In practice the two rafters numbered 1—5 and 5—9 are continuous; similarly the ties numbered 1—12, 9—11, 5—12 and 5—11 are also continuous; whilst the only short members are those forming the internal triangulation, *i.e.* 2—13, 3—13, *etc.* From the point of view of fabrication this produces the least number of separate members and, in consequence, a reduction in the amount of jointing.

89

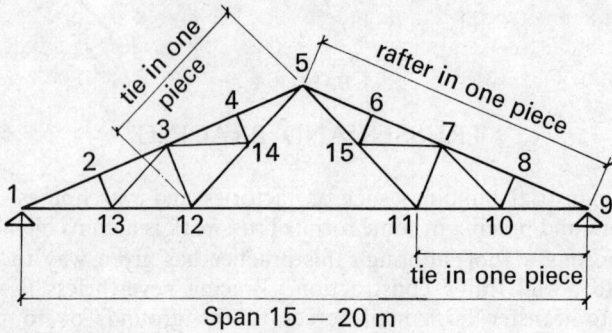

Span 15 — 20 m

FIG. 37

Although it is assumed in analysing the truss that each member is pin jointed at the nodes, this does not, in fact, occur in practice, but convention has it this way and allows the use of statics to arrive at the member forces.

Similarly, it is also assumed that applied loads act at the nodes. Now a fairly high proportion of the loading is transmitted to the truss through the purlins which usually do not coincide with the node points. This is because the spacing of purlins is controlled primarily by the type of roof cladding and secondly by the span of the truss rather than its triangulation. Purlin loads then produce bending moments in the rafter and, because rafters are continuous, they will act as continuous beams subjected to an axial force.

A truss resting on columns will act as a prop against horizontal wind forces and will ensure that both columns deflect by the same amount. As a result additional forces will be introduced into the truss.

When the truss is detailed it will be found that the rafters will require about three bolts at each end (joints 1, 5, 9), the ties a similar number, and all the internal members about two bolts. If these bolting requirements are taken together with the necessary size of the gussets any semblance to a pin-jointed framework disappears completely. Instead of bolting, all joints may be welded and the gussets omitted, however, welding infers a degree of rigidity which suggests an indeterminate system.

It is not intended from the above to suggest that a truss cannot be designed. Provided all these factors are taken into consideration, it is possible to develop a design which is satisfactory in service. Generally these remarks apply equally to bracing systems.

LOADING

There are three different types of load;
1. *Dead*—due to self-weight, weight of purlins and cladding
2. *Imposed*—e.g. loading due to snow
3. *Wind*

Each different type of loading should be computed separately, and set out in tabular form showing whether tensile or compressive. These types of loading may then be combined to give a number of loading conditions as follows:

1. dead and imposed;
2. dead, imposed and wind;
3. dead and wind.

Usually there are two possible wind conditions so that conditions 2 and 3 will produce five different loading situations to be considerered in design. In condition 1 dead and imposed loads normally act in the same direction. However in conditions 2 and 3 the effect of wind may either be to increase or decrease the force in the members and sometimes cause a reversal of forces.

It cannot be overstated that the assessment of the loads should be as accurate as possible. Whilst imposed and wind loads are governed by C.P. 3 Chap. V. an intelligent guess as to the dead load is initially inevitable, but this will improve with experience.

From the initial assessment of the loading a first design may be worked out. Using the self weights of the members obtained from the first design a more accurate estimate of the loading may be made and compared with the original. If necessary the design calculations should be reworked.

FACTORS INFLUENCING THE DESIGN OF MEMBERS

Each member must be designed for the maximum force to which it is subjected having regard to the loading cases previously mentioned. The approach to the design of the members is to assume that condition 1 (dead and imposed) is the criterion, and compare this with the other load cases.

If condition 2 (dead, imposed and wind) produce forces which are less than condition 1 or not more than 25 per cent in excess then it may be ignored since the increase in allowable stress because of wind will not cause a change in member size. Condition 2 becomes the criterion for member design if the forces produced in a given member exceed load case 1 by more than 25 per cent.

Similarly if condition 3 (dead and wind) produces a force reversal in any member then the size of that member must be checked for both load cases. There will be occasions when condition 2 produces force reversals, however condition 3 will be the more significant. It is important to appreciate that where a tie under condition 1 becomes a strut under conditions 2 or 3 the member size may increase even if the strut force is smaller.

Problems occur if the main horizontal ties in Fig. 37 become struts. The longest unsupported length is between node points 11—12 where in–plane restraint is provided, but the out-of-plane unrestrained length is the truss span. Where this condition obtains it will be necessary to provide stabilising members in the plane of the roof. A simple method would be to connect adjacent trusses together along node lines 11 and 12 with light angle ties, which should be capable of resisting a force of not less than 2·5 per cent of the force in members they restrain.

It may be necessary to break up the truss for transporting. The truss shown in Fig. 37 would have a site joint at the apex and at nodes 11 and 12. This must be considered in designing the joints.

Welded trusses can look efficient. It is sometimes possible to eliminate the gussets by lapping the members. For this type of construction rafters made from T sections (cut U.B. or U.C.) are preferred to double angles.

Trusses fabricated from tubes should also be considered. Tubular trusses are usually of welded construction.

CALCULATION OF SIZES OF MEMBERS

Having regard to the remarks in the previous section, member sizes may be computed using the procedures given in Chapter 3. However, the following additional points should be noted.

The rafters are restrained in two directions: firstly by the internal members in the vertical plane, and secondly, by the purlins in the horizontal plane. As previously explained the purlins do not always coincide with the nodes, so that there will be two effective lengths and therefore two slenderness ratios. It is the greater slenderness ratio which controls the allowable stress. Where the purlin points do not coincide with the node points the rafter may be treated as a continuous beam and analysed by moment distribution methods treating the first (eaves) and the last (apex) supports as simply jointed.

Where a tie is subject to a reversal of force due to wind, its slenderness ratio as a strut should not exceed 350. A limit of 180 applies to the slenderness ratio of all struts unless the force is due to wind only when it may be increased to 250.

In analysis it is assumed that the forces in each member meet at a node point and thus produce no secondary effects. However, during detailing the members will be set out on their bolt lines or may be repositioned to improve gusset details, thus introducing secondary effects which should be taken into account in sizing members.

For small trusses single angles will be adequate for both ties and struts. As span and loading increase it is practical to use double angles for rafters and principal ties. Usually the internal members of simple trusses will be single angles. As the load increases double angles produce a more economical use of material and also improve lateral stiffness. For very large trusses double channels will be more economical than double angles. If the trusses are to be bolted, grade 4·6 black bolts are sufficient with a practical minimum size of 16 mm diameter. The holes should be taken into account in ties.

EXAMPLE 25

Fig. 38 shows the arrangement of a braced bay in the side of a factory. The forces are due to wind and are reversible. Design the diagonal bracing members and ensure that the resulting horizontal deflection at the top of the columns does not exceed $h/325$. Assume the columns are $251 \times 146 \times 43$ U.B. and the eaves beam is $203 \times 133 \times 25$ U.B.

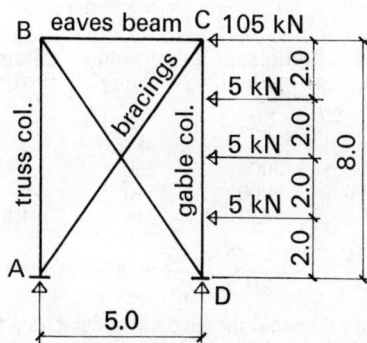

FIG. 38

SOLUTION

There are three possible approaches to the design of this bracing:

1. One member only acting, *i.e.* design for tension
2. One member only acting, *i.e.* design for compression
3. Both members acting, *i.e.* design for compression

Each case will be treated separately to show the effect of these approaches.

Case 1

For the direction of the forces shown the brace AC is ignored and brace BD resists the resultant wind force in tension.

Total force at C = $105 + (0.5 \times 3 \times 5) = 112.5$ kN

Length of brace = $\sqrt{8^2 + 5^2} = 9.43$ m

$\therefore$ Force in brace = $\dfrac{112.5 \times 9.43}{5} = 212.3$ kN

For grade 43 steel permissible tensile stress p_c + wind is $155 + 25\%$

$$= 193.75 \text{ N/mm}^2$$

Hence the net or effective area required,

$$A_n = \frac{212.3 \times 10^3}{193.75} = 1095.7 \text{ mm}^2$$

try a single $100 \times 65 \times 10$ angle long leg attached by 20 mm diameter bolts

Area of attached leg A_1 = $10(95 - 22) = 770$ mm^2
Area of outstanding leg $A_2 = 10 \times 60 = 600$ mm^2

$$\text{Factor, } k = \frac{3 \times 770}{3 \times 770 + 600} = 0\cdot79$$

So that available net area,

$$A_n = 770 + 0\cdot79 \times 600 = 1246\cdot3 \text{ mm}^2$$

As this is greater than required net area angle is satisfactory for strength.
Checking for horizontal deflection, the allowable movement is limited to

$$\delta_a = \frac{8000}{325} = 24\cdot6 \text{ mm}$$

The actual horizontal deflection at C of the column CD may be obtained using the unit load strain energy method. Applying a unit force $k = 1$ horizontal at C gives in tabular form

Member	F	l	A	k	Flk/A
CB	$-112\cdot5$	5000	3230	$-1\cdot00$	174.15
AB	$-180\cdot0$	8000	5510	$-1\cdot60$	418.15
BD	$212\cdot6$	9430	1246	$1\cdot89$	3041.02

so that $\delta = \dfrac{1}{E} \sum \dfrac{Flk}{A} = \dfrac{3633\cdot32 \times 10^3}{210 \times 10^3} = 17\cdot3$ mm

Since this is less than allowable the angle is satisfactory for stiffness.
Use one $100 \times 65 \times 10$ angle.

Case 2

For the direction of the forces shown the brace BD is ignored and the brace AC resists the resultant wind force in compression.
The force in the brace will be $212\cdot3$ kN as previously calculated, but compressive.

Try two $150 \times 150 \times 10$ angles, area $= 58\cdot5$ cm^2, $r = 4\cdot62$ cm (min)

$$\frac{l}{r} = \frac{0\cdot85 \times 9430}{46\cdot2} = 173\cdot5$$

for grade 43 steel $p_c = 31$N/mm$^2 + 25\%$ wind allowance
so that safe load $= 31 \times 1\cdot25 \times 5850 \times 10^{-3} = 226\cdot7$ kN
which is adequate.

Employing the deflection procedure previously mentioned, it will be seen that member CB and AB are now neutral and the force in CD has the same Flk/A value as AB in Case 4. Member AC produces an Flk/A value of $647\cdot7$. The resulting horizontal deflection is $5\cdot08$ mm.
Using two $150 \times 150 \times 10$ angles.

Case 3

With both braces acting jointly the panel will be statically indeterminate. However, it may be assumed that for a single panel of bracing as shown in Fig. 38 the force in each brace can be obtained by taking half the force acting at C and resolving as given in Case 1. This is illustrated in Fig. 39.

From Fig. 39 it will be seen that for the direction of the force at C member AC is compressive whilst member BD is tensile. When the force at C reverses, the forces in members AC and BD will have the same values but their signs will reverse. Hence the criterion for design is compression. Design for $106\cdot13$ kN compression.

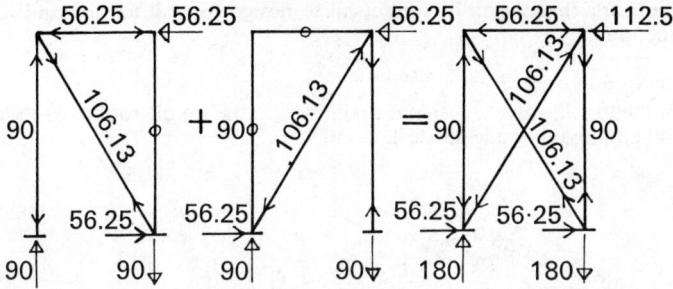

FIG. 39

Try two 120 × 120 × 10 angles. Area = 46·4 cm², r = 3·67 cm (min)

$$\frac{l}{r} = \frac{0·85 \times 9430}{36·7} = 218·4 \ (<250)$$

for grade 43 steel p_c = 20N/mm² + 25% wind allowance
so that safe load = 20 × 1·25 × 4640 × 10⁻³ = 116 kN,
which is just adequate.

The horizontal displacement at C is found as for case 1:

Member	F	l	A	k	Flk/A
DC	90·00	8000	5510	0·800	104·54
CB	−56·25	5000	3230	−0.500	43·54
BA	−90·00	8000	5510	−0·800	104·54
BD	106·13	9430	4640	0·943	203·40
CA	−106.13	9430	4640	−0·943	203·40
					659·42

so that

$$\delta = \frac{659·42 \times 10^3}{210 \times 10^3} = 3·14 \text{ mm}$$

and since this displacement is less than that permitted *i.e.* 24·6 mm the bracing chosen is satisfactory for stiffness.

The total mass of bracing required for each approach is as follows:

Case (1) 2 angles at 12·3 kg/m by 9·43 m long = 232 kg
Case (2) 4 angles at 23·0 kg/m by 9/43 m long = 867 kg
Case (3) 4 angles at 18·2 kg/m by 9·43 m long = 686 kg

Clearly the tension-only approach is the most economical in its use of material, and is generally the procedure adopted in simple design.

In justifying the tension only approach it will be seen that a single 100 × 65 × 10 angle, after allowing for an effective length factor of 0·85, will give a slenderness ratio of nearly 577. This exceeds the limit of 350 allowed in B.S. 449 for tension members acting as struts. For the calculated l/r value the corresponding permissible stress would be 2·92 N/mm², which means that the safe load as a strut would be only 5·7 kN inclusive of wind allowance. Referring to Fig. 38 it will be seen that whilst the force at C is very small both braces AC and BD will act. When the force at C is large enough brace AC will reach its Euler load, about 9·7 kN, and buckle out of plane. Since no further load can be resisted the subsequent increase in the applied force at C will be transferred through the eaves beam and into the tensile brace BD. The Euler load of 9·7 kN represents about 5 per cent of the total brace force and for this reason it is practical to design for full tension and ignore strut action.

EXAMPLE 26

For the loading given, design the members in the simple roof truss shown in Fig. 40 employing grade 43 steel.

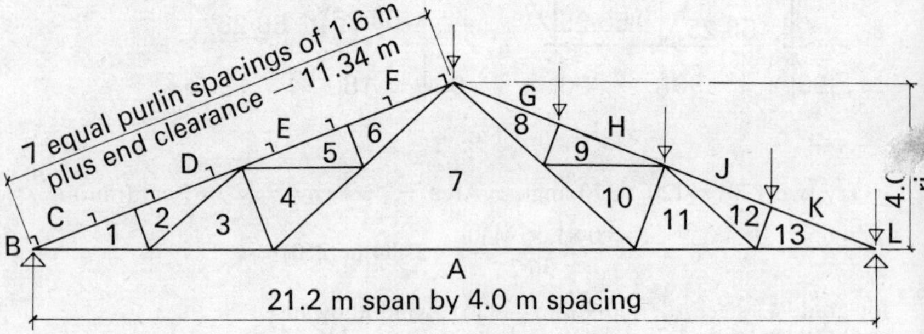

FIG. 40

Loading:

imposed loading from C.P. 3 Chap. V	0·75 kN/m²	
self weight roof cladding	0·27	,,
self weight purlins	0·08	,,
self weight truss	0·10	,,

All loading measured on plan area

SOLUTION

1. Calculate the total dead and imposed load and distribute between the rafter node points
2. Separately calculate the loads on each purlin required for bending effects
3. Determine forces in all members and set out in tabular form, identifying members by Bow's notation
4. Design members.

Horizontal distance between rafter node points is $21 \cdot 2 \div 8 = 2 \cdot 65$ m
Total dead load per unit area $= 0 \cdot 27 + 0 \cdot 08 + 0 \cdot 10 = 0 \cdot 45$ kN/m²
Dead load per intermediate node $= 0 \cdot 45 \times 4 \times 2 \cdot 65 = 4 \cdot 8$ kN
Dead load per end node $= 0 \cdot 45 \times 4 \times 1 \cdot 33 = 2 \cdot 4$ kN
Ditto imposed load   intermediate $= 8 \cdot 0$ kN
               ends $= 4 \cdot 0$ kN

Produce a force diagram for the dead loads (only half of the truss need be considered as it is symmetrical). The imposed load forces may be obtained pro-rata. See Fig. 41 and table.

It is assumed that these loads may be applied at the node points of the rafter (this is the most convenient method of approach), and provided the graphics are prepared accurately to a reasonable scale, the resulting forces may be scaled to 0·1 of a kilonewton, which is close enough for practical design.

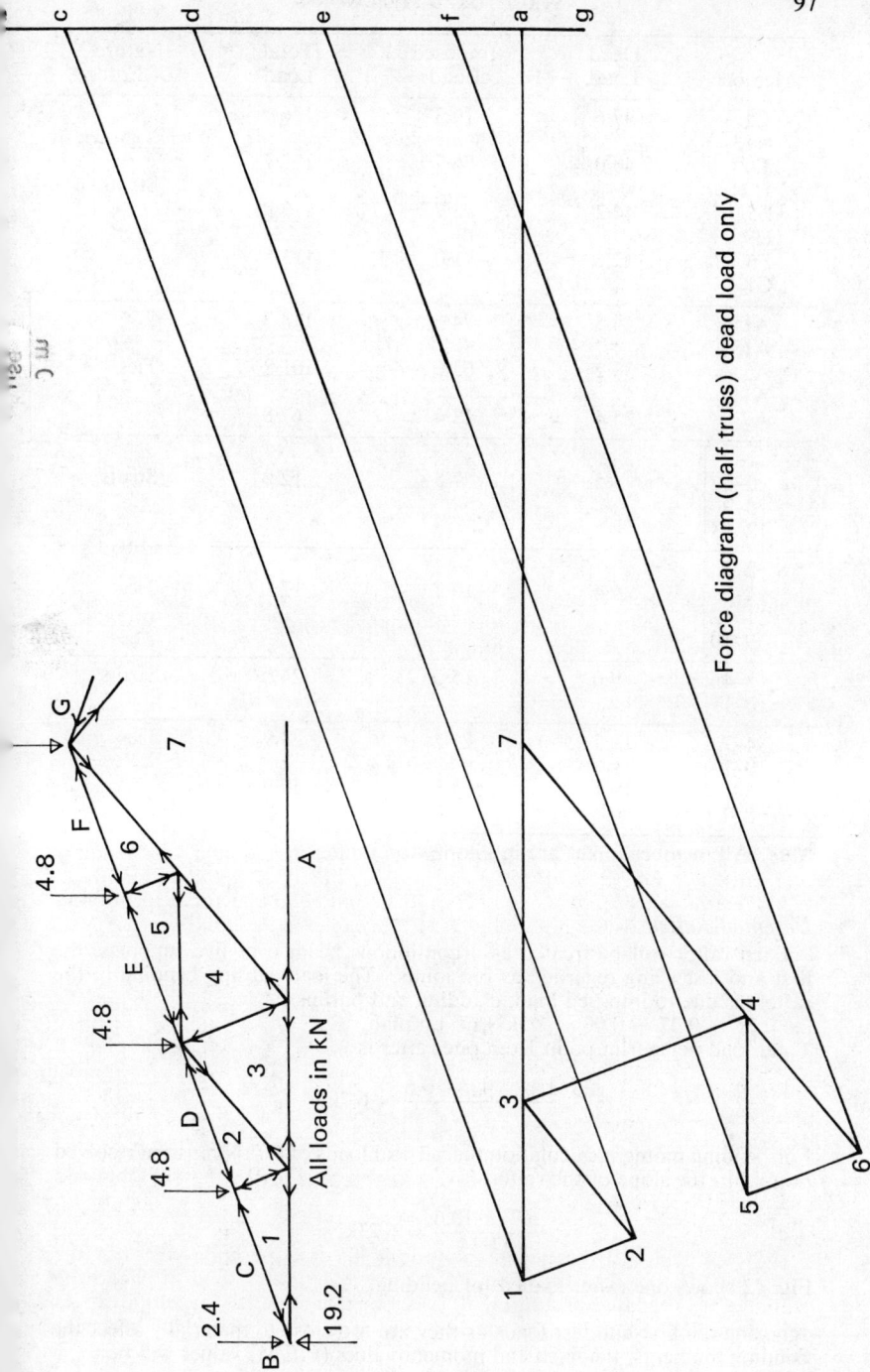

All loads in kN

Force diagram (half truss) dead load only

FIG. 41

Member	Dead Load	Imposed Load	Total Load	Nature of Force
C1 K13	47·6	79·3	126·9	
D2 J12	46·0	76·7	122·7	Struts
E5 H9	44·2	73·7	117·9	
F6 G8	42·6	71·0	113·6	
A1 A13	44·5	74·2	118·7	
A3 A11	38·2	63·6	101·8	Ties
A7	25·4	42·4	67·8	
1–2 5–6 8–9 12–13	4·5	7·5	12·0	Struts
2–3 4–5 11–12 9–10	6·4	10·7	17·1	Ties
3–4 10–11	9·0	15·0	24·0	Struts
4–7 10–7	12·7	21·2	33·9	Ties
6–7 8–7	19·1	31·8	50·9	

Note: All member forces are in kilonewton units.

Design of Rafter

Each rafter will be treated as a continuous beam over five supports, the first and last being regarded as pin joints. The load causing bending in the rafters is due to imposed load, cladding and purlins.
i.e. $0.75 + 0.27 + 0.08 = 1.1 \text{kN/m}^2$ on plan.
Total load per purlin point from one rafter is

$$\frac{1.1 \times 4 \times 10.6}{7} = 6.7 \text{kN}$$

For bending moment calculation the vertical loads of 6·7kN must be resolved normal to the slope of the rafter.

$$\frac{6.7 \times 10.6}{11.34} = 6.23 \text{kN}$$

Fig. 42 shows one rafter loaded for bending

Ignoring the first and last loads as they are too close to materially affect the bending moments, the fixed end moment values (F.E.M) values will be

For BA and DE, F.E.M. $= \dfrac{6 \cdot 23 \times 1 \cdot 67 \times 1 \cdot 16}{2 \cdot 83^2}\left(1 \cdot 67 + \dfrac{1 \cdot 16}{2}\right) = 3 \cdot 39$ kNm

For BC and DC, F.E.M. $= \dfrac{6 \cdot 23 \times 0 \cdot 44 \times 2 \cdot 4^2}{2 \cdot 84^2} + \dfrac{6 \cdot 23 \times 2 \cdot 04 \times 0 \cdot 8^2}{2 \cdot 84^2}$

$= 2 \cdot 97$ kNm

For CB and CD, F.E.M. $= \dfrac{6 \cdot 23 \times 0 \cdot 44^2 \times 2 \cdot 4}{2 \cdot 84^2} + \dfrac{6 \cdot 23 \times 2 \cdot 04^2 \times 0 \cdot 8}{2 \cdot 84^2}$

$= 2 \cdot 93$ kNm

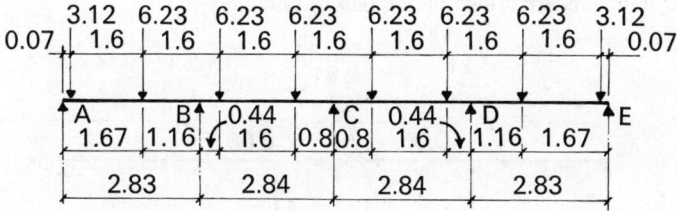

FIG. 42

Since the rafter section is uniform throughout I can be taken as unity and the member stiffnesses (I/L) are:

$$K_{BA} = \frac{0 \cdot 75}{2 \cdot 83} = 0 \cdot 265 \text{ and } K_{DE}$$

$$K_{BC} = \frac{1}{2 \cdot 84} = 0 \cdot 352 \text{ and } K_{CD}$$

Making the distribution factors

$$d_{BA} = \frac{0 \cdot 265}{0 \cdot 265 + 0 \cdot 352} = 0 \cdot 43$$

$$d_{BC} = \frac{0 \cdot 353}{0 \cdot 265 + 0 \cdot 352} = 0 \cdot 57$$

$$d_{CB} = d_{CD} = 0 \cdot 5$$

The final joint moments are shown in the following table:

Distribution factors	0·43	0·57	0·5	0·5	0·57	0·43
F.E.M.	+3·39	−2·97	+2·93	−2·93	+2·97	−3·39
Out of balance moment	−0·18	−0·24	0	0	+0·24	+0·18
Carry over moments			−0·12	+0·12		
			0	0		
Final moment	+3·21	−3·21	+2·81	−2·81	+3·21	−3·21

Simple span moments are

$$AB, \qquad M = \frac{6{\cdot}23 \times 1{\cdot}67 \times 1{\cdot}16}{2{\cdot}83} = 4{\cdot}26 \text{ kNm}$$

$$BC \text{ (nearest } B\text{)}, \; M = \frac{6{\cdot}23 \times 0{\cdot}44}{2{\cdot}84} (1{\cdot}6 + 2 \times 0{\cdot}8) = 3{\cdot}09 \text{ kNm}$$

$$\text{(nearest } C\text{)}, \; M = \frac{6{\cdot}23 \times 0{\cdot}8}{2{\cdot}84} (2 \times 0{\cdot}44 + 1{\cdot}6) = 4{\cdot}35 \text{ kNm}$$

The final bending moments in the rafter are as shown in Fig. 43. *Note:* moments are drawn on the tension side.

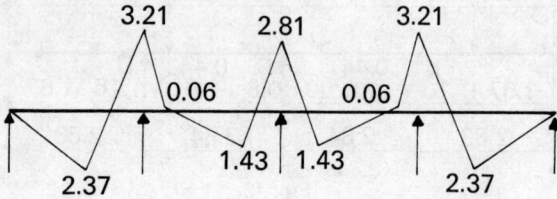

FIG. 43

From the table of forces, rafter member $C1$ carries the greatest direct load of $126{\cdot}9$ kN whilst the span and support moments are $2{\cdot}37$ kNm and $3{\cdot}21$ kNm respectively.

Try two $100 \times 75 \times 8$ angles long legs arranged vertically back to back, and separated by 10 mm gussets.

The properties of the two angles compounded are:

$A = 26{\cdot}9$ cm^2, $r_x = 3{\cdot}14$ cm, $r_y = 3{\cdot}22$ cm,
Z_x (top) $85{\cdot}8$ cm^3 (bottom) $38{\cdot}6$ cm^3, $c_x = 3{\cdot}1$ cm

For the permissible compressive stress p_c two slenderness ratios must be considered, the greater of these values should be used.

About the xx axis the l/r value will be related to the distance between the truss node points and an effective length factor of $0{\cdot}85$ may be taken, this gives

$$\frac{l}{r_x} = \frac{0{\cdot}85 \times 2830}{31{\cdot}4} = 76{\cdot}6$$

About the yy axis the l/r value will be measured between the purlin positions, so that the effective length is equal to the actual length.

$$\frac{l}{r_y} = \frac{1600}{32{\cdot}2} = 49{\cdot}7$$

The xx axis slenderness controls giving $p_c = 107$ N/mm (Table 17a)

$$f_c = \frac{126{\cdot}9 \times 10^3}{26{\cdot}9 \times 10^2} = 47{\cdot}2 \text{ N/mm}^2$$

and

$$\frac{f_c}{p_c} = \frac{47{\cdot}2}{107} = 0{\cdot}44$$

Since this ratio is significantly less than unity the trial section is reasonable. Check for bending.

In this case the slenderness ratio about the yy axis applies.

The maximum bending moment is at the first internal support and this will produce compressive stress in the downstanding legs of the rafter. According to B.S. 449 the permissible stress should be computed by treating the angles as a case III plate girder in which K_2 is taken as -1 whilst the thickness of the downstanding leg of the angle should be used for the D/T ratio.

Since there are two angles connected together, T will be taken as the sum of their thicknesses, hence

$$\frac{D}{T} = \frac{100}{2 \times 8} = 6 \cdot 16$$

Referring to Chapter 2 and noting the value given to K_2 the critical stress expression will become

$$C_s = (A - B)\frac{y_c}{y_t}$$

where

$$A = \left(\frac{1675}{49 \cdot 7}\right)^2 \sqrt{1 + \frac{1}{20}\left(\frac{49 \cdot 7}{6 \cdot 16}\right)^2} = 2343$$

and

$$B = \left(\frac{1675}{49 \cdot 7}\right)^2 = 1136$$

whilst $y_c = d - c_x = 100 - 31 = 69$
and $y_t = c_x = 31$

Hence $C_s = (2343 - 1136)\dfrac{69}{31} = 2687 \text{ N/mm}^2$

and from Table 8 (B.S. 449), $p_{bc} = 165 \text{ N/mm}^2$

now $f_{bc} = \dfrac{3 \cdot 21 \times 10^6}{38 \cdot 6 \times 10^3} = 83 \cdot 2 \text{ N/mm}^2$

and $\dfrac{f_{bc}}{p_{bc}} = \dfrac{83 \cdot 2}{165} = 0 \cdot 5$

giving $\dfrac{f_c}{p_c} + \dfrac{f_{bc}}{p_{bc}} = 0 \cdot 44 + 0 \cdot 5 = 0 \cdot 94 < 1$

Thus the proposed rafter size is satisfactory.

Suitably connected spacer packs must be provided along the length of the rafter to control the slenderness ratio in each separate angle. This is to ensure that each angle will be capable of generating the permissible stress, p_c. Since $\dfrac{l}{r_x} > \dfrac{l}{r_y}$ the former is the least favourable, and the spacing should be limited to a local slenderness ratio of 40 or $0 \cdot 6 \times 76 \cdot 6$ whichever is the smaller. In this instance 40 governs.

For a single $100 \times 75 \times 8$ angle the least radius of gyration is 16 mm taken about the vv axis. This gives a maximum spacing of $40 \times 16 = 640$ mm. At least three connections should be made between nodes.

Joints

There are five bolted joints along each rafter.

1. The eaves joint force equals $126 \cdot 9$ kN ($C1$) and the bolts required here will be in double shear.

Using 20 mm diameter bolts grade 4·6 and 10 mm gussets,

Double shear value $= \dfrac{2 \times 80 \times 20^2 \times \pi \times 10^{-3}}{4} = 50\cdot3$ kN

Bearing value $= 250 \times 20 \times 10 \times 10^{-3} = 50$ kN

Number of bolts required $= \dfrac{126\cdot9}{50} = 2\cdot54$ say 3

2. Apex joint force equals 113·6 kN (F6), using 20 mm diameter bolts:

Number of bolts required $= \dfrac{113\cdot6}{50} = 2\cdot27$, say 3

3. Intermediate joints

Since the rafter is unbroken at these joints only the force difference need be considered:

Member $C1$ to $D2$:   net bolt force $= 126\cdot9 - 122\cdot7 = 4\cdot2$ kN
Member $D2$ to $E5$:   net bolt force $= 122\cdot7 - 117\cdot9 = 4\cdot6$ kN
Member $E5$ to $F6$:   net bolt force $= 117\cdot9 - 113\cdot6 = 4\cdot3$ kN

Using 16 mm dia. bolts, double shear value $= 32$ kN and bearing value $= 40$ kN. Provide at least two bolts each.

Design of Struts.
Members 1–2 : 5–6 : 8–9 : 12–13. Load 12 kN.

As this force is very small try a single angle, *e.g.* 60 × 30 × 5 long leg connected with two bolts each end. $A = 4\cdot29$ cm², $r_{vv} = 0\cdot63$ cm

Length between intersections is 1·07 m. An effective length factor of 0·85 may be used as both ends are double bolted.

$$\frac{l}{r} = \frac{0\cdot85 \times 1070}{6\cdot5} = 144\cdot4$$

from Table 17a $p_c = 43$ N/mm²
∴ safe load $= 43 \times 429 \times 10^{-3} = 18\cdot4$ kN > 12

Members 3,4 : 10,11. Load 24 kN.

Try a single angle 80 × 60 × 6 long leg connected and double bolted
$A = 8\cdot11$ cm² $r_{vv} = 1\cdot29$ cm

Length between intersections $= 2\cdot14$ m. Effective length factor 0·85

$$\frac{l}{r} = \frac{0\cdot85 \times 2140}{12\cdot9} = 141$$

from Table 17a $p_c = 45$ N/mm²
∴ safe load $= 45 \times 811 \times 10^{-3} = 36\cdot5$ kN > 24
Note: both a 75 × 50 × 8 and 65 × 50 × 8 are satisfactory and would produce a more efficient stress ratio, but are slightly heavier.

Design of Ties.
Members A1, A13, A3, A11. Load 118·7 kN (A1).

Members A1 and A3 will be in one length, as will A11 and A13.

Try a pair of angles 65 × 50 × 5 long leg connected. Gross Area = 11·08 cm
Employ 20 mm diameter bolts

Net area required = $\dfrac{118\cdot7 \times 10^3}{155}$ = 765·8 mm²

Net area provided = 1108 − (2 × 22 × 5) = 888 mm²

Note: A pair of 60 × 30 × 6 angles would be safe, but only 16 mm diameter
bolts would be possible. Since the number of bolts would increase so would
the workmanship content. Equally a single 80 × 60 × 8 angle would be
adequate, but twice the number of bolts would be required since the
connections will be in single shear.

Use two 65 × 50 × 5 angles.

Member A7.

Load 67·8 kN

Since this member is over 9 m long a sag bar will be used. This will be connected to the apex gusset and at the centre of the tie. A nominal 50 × 6 flat
will be adequate for this purpose.
Try a pair of angles 60 × 30 × 5 long leg connected
Gross area = 8·58 cm, employ 16 mm diameter bolts.

Net area required = $\dfrac{67\cdot8 \times 10^3}{155}$ = 437·4 mm²

Net area provided = 858 − (2 × 18 × 5) = 678 mm²

This is more than adequate, but is consistent with the previous panels A1, etc.
Members 4,7 : 10,7 : 6,7 : 8,7.

Load 50·9 kN

Members 4,7 and 6,7 would be in one length as would 8,7 and 10,7.

Try a single angle 65 × 50 × 5 long leg connected and employ 20 mm
diameter bolts.

Gross area = 5·54 cm²

Net area required $\qquad\qquad$ = $\dfrac{50\cdot9 \times 10^3}{155}$ = 328·4 mm²

Net area of connected leg $\quad$ = (62·5 − 22) 5 = 202·5 mm²

Net area of outstanding leg = 47·5 × 5 = 237·5 mm²

Reduction factor k $\qquad\qquad$ = $\dfrac{3 \times 202\cdot5}{(3 \times 202\cdot5) + 237\cdot5}$ = 0·72

Net area provided = 202·5 − (0·72 × 237·5) = 373·5 mm², which is
adequate

Members 2,3 : 4,5 : 11,12 : 9,10.

Load 17·1 kN

Try a single angle 60 × 30 × 5 long leg connected and employ 16 mm diameter bolts.

Gross area = 4·29 cm²

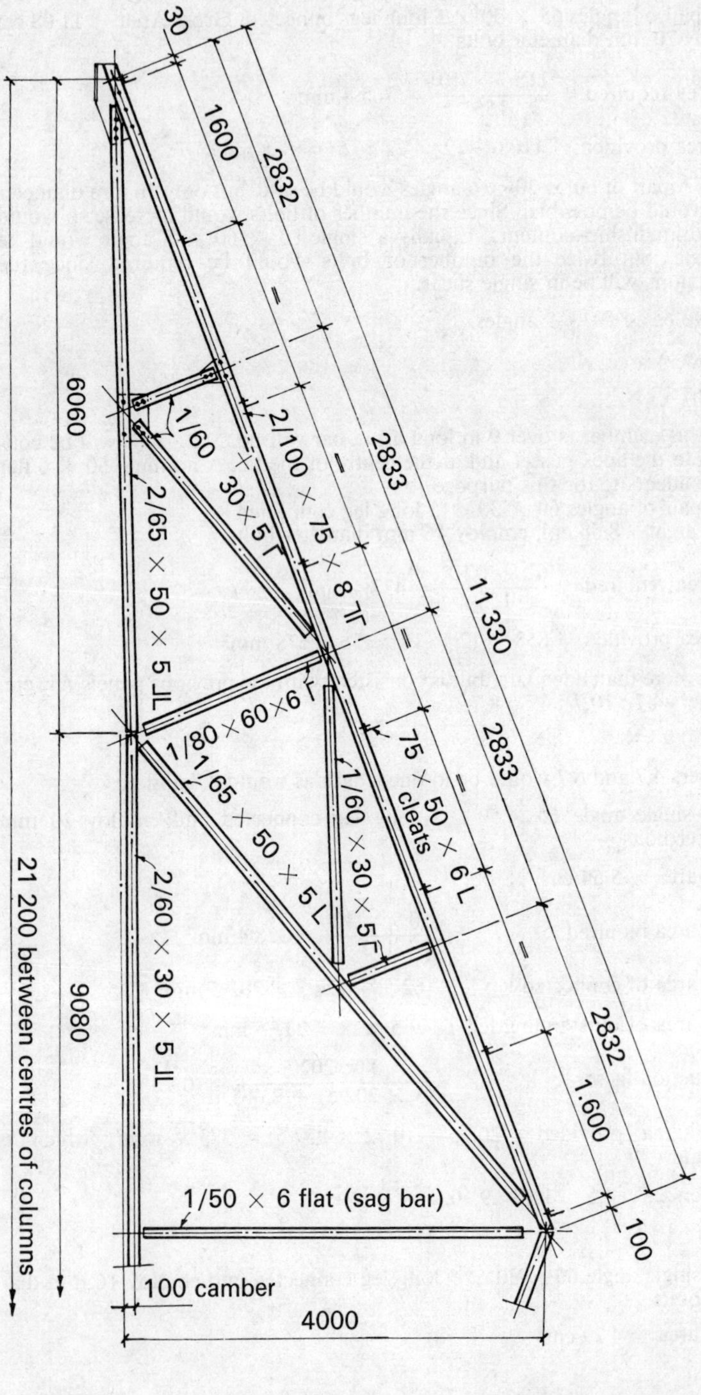

30
1600
2832
2/100 × 75 × 8 ⅃L
1/60
30 × 5 Γ
2833
11 330
2/65 × 50 × 5 ⅃L
1/80 × 60 × 6 Γ
1/65 × 50 × 5 L
1/60 × 30 × 5 Γ
75 × 50 × 6 L
cleats
2833
2832
1.600
100
6060
9080
2/60 × 30 × 5 ⅃L
21 200 between centres of columns
1/50 × 6 flat (sag bar)
100 camber
4000

FIG. 44

Net area required $= \dfrac{17 \cdot 1 \times 10^3}{155} = 110 \cdot 3 \text{ mm}^2$

Net area of connected leg $= (57 \cdot 5 - 18)\, 5 = 197 \cdot 5 \text{ mm}^2$

Net area of outstanding leg $= 27 \cdot 5 \times 5 = 137 \cdot 5 \text{ mm}^2$

Reduction factor k $= \dfrac{3 \times 197 \cdot 5}{(3 \times 197 \cdot 5) + 137 \cdot 5} = 0.81$

Net area provided $= 197 \cdot 5 + (0 \cdot 81 \times 137 \cdot 5) = 309 \cdot 1 \text{ mm}^2,$

which is more than adequate

Note: Whilst this member is much stronger than needs be, it is the smallest tested angle given in section tables. The procedure for calculating the number of joint bolts at each end of a member is as shown for the rafter. It is left to the reader to calculate the number of joint bolts required in the remaining members.

To improve the appearance a small amount of camber will be built in; this will remove the sagging effect caused by deflection.

CHAPTER 5

PLASTIC THEORY AND DESIGN

THIS chapter is concerned with simple plastic theory and its application to the design of the elements of a steel-framed structure.

Plastic theory should not be regarded as a universally alternative method to elastic design, but should only be considered where the structure under review is amenable to its principles. The main difference between the elastic and plastic design methods is briefly as follows.

An elastically designed structure employs the working or service loading in assessing the stresses in any component member, and these stresses are limited and remain within the elastic strength of the steel used.

A plastically designed structure employs the working or service loading increased by a suitable load factor and allows the proportions of any component member to be assessed using the yield stress of the steel. Thus under factored loading the strength of the structure is just at its collapse value.

In general terms instability should not be permitted to precede plastic collapse. Expressing this requirement from the standpoint of elastic design, no structural framework should be considered for plastic design if the maximum permissible stresses from B.S. 449 suffer reduction on account of lack of restraint.

The conditions upon which the simple plastic theory is based may be summarised as follows:

1. instability is prevented;
2. the steel is ductile and complies with B.S. 4360;
3. the major contribution to plastic action comes from bending, and shear and axial faces are of a lower order;
4. strength is the essential criterion and deflection is subservient.

BASIC THEORY

It will be recalled that in the simple tensile test the load extension or stress-strain relationship produces a curve of the form shown in Fig. 45. If, instead, a beam is subjected to a bending test then load-deflection or moment-curvature relationships may be obtained and these would produce curves similar to those already mentioned and shown in Figs. 45(a)—(c).

Now imagine that Fig. 45 (*a*) represents the bending stress-strain relationship (which it does), then all three diagrams are connected and convey the history of the bending test. Within the context of the meaning of simple plastic theory it would be observed that from this beam test:

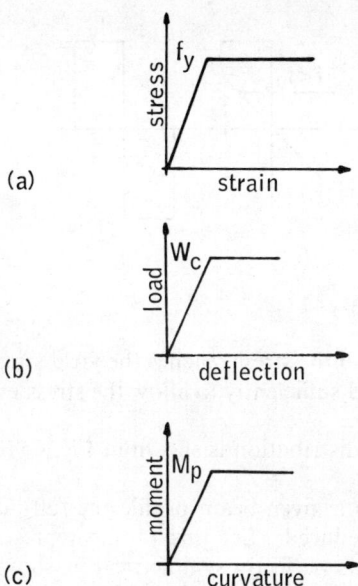

FIG. 45

(*a*) extensive straining occurs without increase of stress beyond the value f_y (Fig. 45 (*a*));

(*b*) large deflections would result without any corresponding increase in load above W_c (Fig. 45 (*b*));

(*c*) curvature would also increase without further increase in the moment above the value of M_p (Fig. 45 (*c*)).

Hence, it may be concluded that:

(*a*) the maximum stress which may be induced is the yield stress of the material f_y;

(*b*) the maximum load which can be supported is represented by W_c which may be defined as the collapse load; and

(*c*) the corresponding maximum bending moment which can be resisted may be defined as the full plastic moment M_p.

SECTIONAL PROPERTIES—1

When the applied bending moment is very low the induced stress may be found from the simple bending theory (*see* Chapter 2). This stress has a maximum value at the extreme outer fibres and varies uniformly to zero at the neutral axis. The bending moment may be increased and the maximum stress allowed to rise just up to the yield stress f_y; the theory of elastic bending remaining intact. A further increase in the bending moment causes the extreme outer fibres to yield, but the stress remains constant at f_y. Other fibres nearer the neutral axis receive

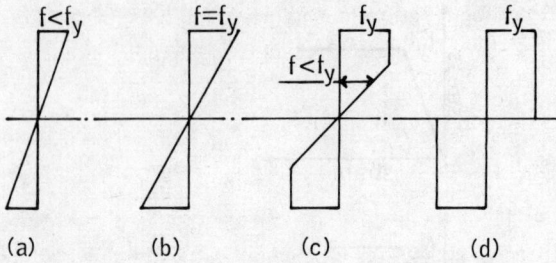

FIG. 46

proportional increases in strain and the stress reaches the yield stress f_y. Eventually all fibres have strained sufficiently to allow the stress everywhere to reach the yield stress f_y .

The development of this stress distribution is shown in Fig. 46 (a) to (d).

From Fig. 46 (d) in relation to a given beam profile the full plastic moment of resistance may be deduced. Let this profile represent a typical universal beam regarded as perfectly symmetrical as shown in Fig. 47.

Now for equilibrium the forces above and below the neutral plane, in the absence of any axial load, must be equal. The moment of these forces measured about say the neutral plane, will be the moment of resistance of the section and in this case the full plastic moment M_p. Setting the areas above and below this neutral plane as A_c and A_t and putting y_c and y_t as the distances from this plane to their respective centroids of force, then since f_y is constant

$$M_p = f_y(A_c y_c + A_t y_t)$$

It will be seen that the quantity within the bracket is the first moment of area of the entire section profile, and is defined as the plastic modulus of the section Z_p, hence

$$M_p = f_y Z_p$$

Z_p is not equal to the elastic section modulus $Z = I/y$ but is related to it. If any given profile is examined and both Z_p and Z calculated it will

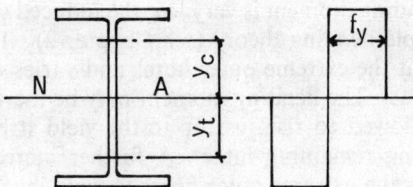

FIG. 47

be found that their ratio for this profile is constant. This ratio (or constant) is termed the shape factor and is written as

$$k_s = \frac{Z_p}{Z}$$

Hence the full plastic moment of resistance may also be expressed as

$$M_p = f_y k_s Z$$

Structural steelwork section tables give values of both Z_p and Z. For a universal beam or column $k_s = 1 \cdot 15$ approximately.

From the foregoing it will be evident that for perfectly symmetrical sections $A_c = A_t$ and $y_c = y_t$, therefore the neutral plane (axis) must also pass through the centroid of the whole section.

SECTIONAL PROPERTIES—2

Suppose a beam profile is not symmetrical, a T section being a good example, the requirement that the forces above and below the neutral plane must be equal is still valid. Using the previous symbols, obviously

$$f_y A_c = f_y A_t,$$

but since the stress is everywhere uniform

$$A_c = A_t$$

as before. However, the plane separating these areas will not correspond with a plane passing through the centroidal axis of the whole profile. Thus a second plane exists about which both the areas and the forces are equal. This plane may be defined as the equal area or equal force axis. The full plastic moment of resistance M_p is obtained by taking moments of the forces above and below this axis. It will be seen to be identical in form to that previously discussed.

LOAD FACTOR

It was defined that W_c represents the collapse load and would yield the full plastic moment M_p. If a collection of beams having different end conditions (free or fixed) and working load W were first designed elastically and then tested to find the ratio W_c/W the result obtained for each test would not always be identical. In fact, in general, only those beams which were simply supported would produce a constant ratio W_c/W and for these cases would be the lowest value obtained.

From a practical point of view a minimum acceptable and constant load factor is required, and that found for a simply supported beam has been deemed satisfactory. Since the ratio W_c/W represents the load factor N, its actual value may be found by considering a simple beam with a central point load. Elastically,

$$M = \frac{WL}{4} = p_{bc}Z$$

Plastically,

$$M_p = \frac{W_cL}{4} = f_y Z_p = f_y k_s Z$$

From which

$$\frac{M_p}{M} = \frac{W_c}{W} = \frac{f_y}{p_{bc}} \cdot k_s$$

using grade 43 steel, $f_y = 250$ N/mm² and $p_{bc} = 165$ N/mm², and a mean value of k_s found from the section tables gives the load factor N,

$$\therefore N = \frac{M_p}{M} = \frac{W_c}{W} = 1 \cdot 75$$

EFFECT OF SHEAR FORCE

The full plastic moment M_p will be reduced by shear force.

From the beam profile shown in Fig. 47 it will be apparent that the areas above and below the neutral plane could conveniently be broken into their constituent parts, *i.e.* flange and web. Summing up the moments of the forces acting on these parts would give the same plastic moment, that is

$$M_p = f_y \cdot BT(d + T) + \frac{f_y t_w d^2}{4}$$

It is assumed (as in the elastic design) that the shear is resisted by the web. Evidence exists elsewhere to the effect that due to shear the yield stress in the second term of the above expression (the web) should be reduced from f_y to f_y', *i.e.*

$$f_y' = \sqrt{(f_y^2 - 3q^2)}$$

where q is the mean shear stress in the web which must not exceed $f_y/\sqrt{3}$. Hence the reduced plastic moment on account of shear is

$$M_p' = f_y BT(d + T) + f_y \frac{t_w d^2}{4} \sqrt{\left(1 - \frac{3q^2}{f_y^2}\right)}$$

Using a slightly different approach this expression may be written as

$$M_p' = M_p - \text{loss of web capacity,}$$

thus loss of web capacity is the plastic modulus of the web ($t_w d^2/4 = Z_{pw}$) multiplied by the difference in the full and reduced yield stresses, that is $f_y - f_y'$,

$$\therefore \text{ loss of web capacity} = Z_{pw}(f_y - f_y')$$

$$= Z_{pw}(f_y - \sqrt{(f_y^2 - 3q^2)}],$$

$$= f_y Z_{pw}\left[1 - \sqrt{\left(1 - \frac{3q^2}{f_y^2}\right)}\right]$$

hence

$$M_p' = M_p - f_y Z_{pw}\left[1 - \sqrt{\left(1 - 3\frac{q^2}{f_y^2}\right)}\right]$$

EFFECT OF AXIAL FORCE

The effect of axial force diminishes the full plastic moment M_p. Consider a rectangular section bd subject to an axial force P and a moment M. Let P remain constant and allow M to increase gradually until all fibres are at the yield stress under the combined effect of both P and the maximum possible value of M. The build-up of stress to the full value f_y is shown in Fig. 48.

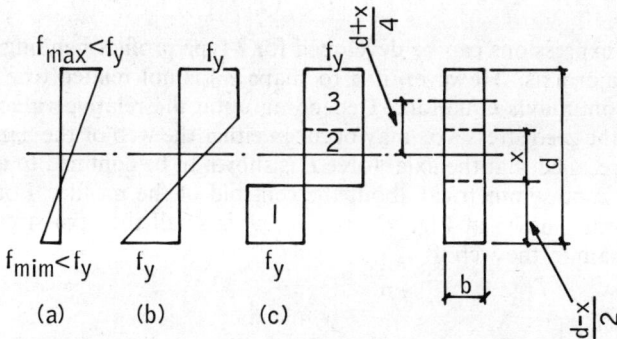

FIG. 48

Initially all stresses are elastic since P and M are small (Fig. 48 (a)). Increasing M, the maximum stress rises to f_y and spreads down the section, and simultaneously f_{min} increases to f_y (Fig. 48 (b)). Further increase in M causes the entire section to reach yield (Fig. 48 (c)). Now this final stress distribution may be divided into three zones, such that zones 1 and 3 represent the effect of the moment M while zone 2 represents the axial force. The axial force is confined to zone 2, such that

$$P = bxf_y$$

Taking moments about the centroid of the entire section

$$M = 2f_y\left[\frac{b}{2}\left(d - x\right)\left(\frac{d + x}{4}\right)\right]$$

and replacing x by βd in both expressions, it will be seen that $P = bdf_y \cdot \beta$ and bdf_y represents the yield load P_y on the section when $M = 0$, then

$$P = P_y\beta$$

Similarly upon simplification the moment expression becomes

$$M = f_y\frac{bd^2}{4}\left(1 - \beta^2\right)$$

Now $f_y \cdot bd^2/4 = M_p$ and substituting for β from $P = P_y\beta$ gives

$$M = M_p\left[1 - \left(\frac{P}{P_y}\right)^2\right]$$

This expression represents the maximum value of the reduced plastic moment in the presence of an axial force for a simple rectangular section. Dividing both sides by the yield stress f_y gives the reduced plastic modulus, that is

$$Z_p' = Z_p\left[1 - \left(\frac{P}{P_y}\right)^2\right]$$

Similar expressions can be developed for I-type profiles bending about their major axis. However, due to shape Z_p' is not related to Z_p by a single continuous equation. Depending upon the relative values of P and M the zero stress axis may occur in either the web or the flange. It will be recalled that the axial force P is shown to be confined to a small area or zone symmetrical about the centroid of the profile. For an I-type profile shown at Fig. 49 provided P is small the zero stress axis will remain in the web if

$$\frac{P}{f_y} = t_w x \leqslant A_w$$

Area of whole section = A
Area of web only = A_W

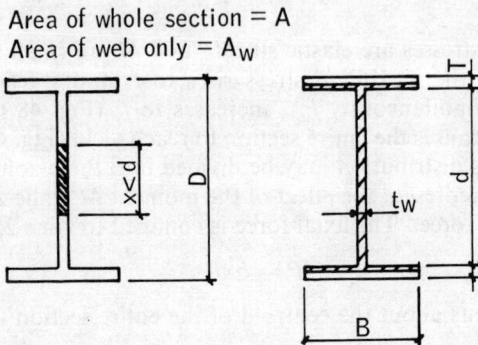

Fig. 49

Hence the value of Z'_p depends upon whether P/f_y is less than or greater than A_w.

If $P/f_y < A_w$ then

$$Z'_p = Z_p - \frac{A^2}{4t_w}\left(\frac{P}{P_y}\right)^2,$$

and if $P/f_y > A_w$ then

$$Z'_p = \frac{A^2}{4B}\left(1 - \frac{P}{P_y}\right)\left(\frac{2DB - A}{A} + \frac{P}{P_y}\right)$$

Steel designers' section tables gives values of Z'_p for both major and minor axis bending, in terms of a coefficient n. In the above expressions

$$n = \frac{P}{P_y}$$

SPECIAL RESTRICTIONS

Earlier it was said that instability must be prevented.

In addition to the need to ensure adequate restraint to enable plastic action to develop, both flanges and web must be stiff enough to prevent premature local buckling. Thus the ratios B/T and d/t_w shown in Fig. 49 are limited. Maximum values are given in the following table.

Steel grade B.S. 4360	B/T	d/t_w	(bending only)
43	18	53	85
50	15·25	44	70
55	13·75	39	65

Depending upon the grade of steel to be used and the limiting ratios given in this table certain universal beams and columns are not suitable for plastic design and others are limited in the amount of axial force which may be resisted. Steel designers' section tables indicate suitability in tabular form and should be consulted.

BEAM ACTION

Consistent with the limitation and restriction and strength provisions given in the preceding sections, plastic design is best suited to beams which are fixed or continuous. Simply supported beams offer no advantage.

When the collapse load has been reached, a sufficient number of plastic hinges must form to create a mechanism. Usually three such hinges are required, one at each fixed end and the other at the position of maximum span moment at collapse. At all these hinges the value of the plastic moment generally must be equal to give a perfect mechanism. Propped cantilevers require two plastic hinges, one at the fixed end and the other in the span; the third hinge is natural, being at the simple support.

The determination of the magnitude of plastic hinge moment may be achieved by work methods or by a "reactant-line" approach. Work methods involve setting up equations relating the external work done by the loads to the internal work done by the hinges in rotating and solving for M_p. The reactant-line approach involves the calculation of the simple bending moments which may be drawn to scale and then superimposing on this diagram a "reactant" bending moment which may be adjusted until the correct M_p is found.

Whichever method is employed it is essential to ensure that the value of M_p determined is nowhere exceeded. In short, for plastic design a structure at collapse must satisfy

1. equilibrium;
2. mechanism;
3. yield.

COLUMN ACTION

In plastic design column behaviour is a complex subject, and that section dealing with the effects of axial load represents only a small part of the total design problem. Nevertheless stability is the main criterion. Briefly the design of a column using plastic methods depends upon:

1. the axial load;
2. the ratio of the terminal moments;
3. major axis slenderness ratio;
4. minor axis slenderness ratio;
5. torsional properties.

A detailed discussion of the dependancies is outside the scope of this book.

CONNECTIONS

The rules given in Chapter 1 generally apply here. However, it must be remembered that design will be related to plastic theory. Connections are best made either by welding or through the use of high-strength, friction-grip bolts. Design stresses will be closely related to yield- or proof-stress values. The working stress approach is generally inappropriate.

EXAMPLE 27
Determine the plastic modulus for the two profiles given in Fig. 50.

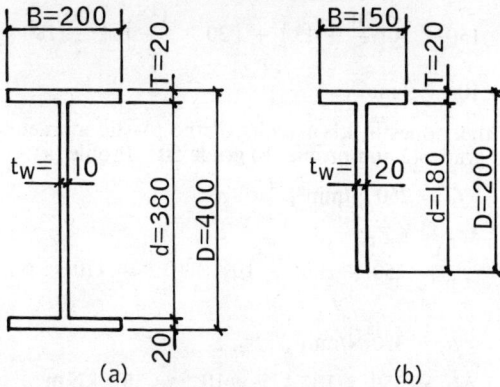

Fig. 50

SOLUTION

Profile (a)

This section is perfectly symmetrical, taking moments about the centroid of the whole area gives

$$Z_p = 2BT\left(\frac{D-T}{2}\right) + 2t_w\left(\frac{d}{2}\right)\left(\frac{d}{4}\right)$$

$$= BT(D-T) + t_w\frac{d^2}{4}$$

$$= 200 \times 20(400 - 20) + \frac{10 \times 380^2}{4}$$

$$= 1881 \times 10^3 \text{ mm}^3$$

Alternatively Z_p may be found by

$$Z_p = \frac{BD^2}{4} - (B - t_w)\frac{d^2}{4}$$

$$= \frac{200 \times 400^2}{4} - (200 - 10) \times \frac{380^2}{4}$$

$$= 1881 \times 10^3 \text{ mm}^3, \text{ as before.}$$

Profile (b)

This section is not symmetrical; find firstly the equal area axis. Measure x downwards from the underside of the flange then

$$BT + t_w x = tw\,(d - x)$$

$$\therefore \quad (150 \times 20) + 20x = 20(180 - x)$$

from which $x = 15$ in the direction assumed.

Hence taking moments about the equal area axis

$$Z_p = BT\left(\frac{T}{2} + x\right) + t_w\frac{x^2}{2} + \frac{t_w}{2}(d-x)^2$$

$$= 150 \times 20\left(\frac{20}{2} + 15\right) + \left(20 \times \frac{15^2}{2}\right) + \frac{20}{2}(180 - 15)^2$$

$$= 104\,475 \text{ mm}^3.$$

What will be the values in kNm units of the plastic moments of resistance if profile (a) is grade 43 and profile (b) grade 50? Profile (a):

$$f_y = 250 \text{ N/mm}^2,$$

$$M_p = f_y Z_p,$$

$$= 250 \times 1881 \times 10^3 \times 10^{-6} = 470.3 \text{ kNm}$$

Profile (b):

$$f_y = 350 \text{ N/mm}^2,$$

$$M_p = 350 \times 104\,475 \times 10^{-6} = 36.6 \text{ kNm}$$

EXAMPLE 28

If the profile shown at Fig. 50 (a) is subject to a factored shearing force of 380 kN calculate the reduced value of the plastic moment of resistance.

SOLUTION

Average shear stress in web is

$$q = \frac{Q}{dt_w} = \frac{380 \times 10^3}{380 \times 10} = 100 \text{ N/mm}^2 < f_y/\sqrt{3};$$

the reduced yield stress in the web is

$$f'_y = \sqrt{(f_y{}^2 - 3q^2)}$$

$$= \sqrt{[250^2 - (3 \times 100^2)]} = 180 \text{ N/mm}^2$$

Z_p for the web is

$$Z_p = \frac{t_w d^2}{4} = \frac{10 \times 380^2}{4} = 361 \times 10^3 \text{ mm}^3$$

then loss of web capacity is

$$Z_{pw}(f_y - f'_y) = 361 \times 10^3(250 - 180) = 25.27 \times 10^6 \text{ Nmm},$$

giving a reduced plastic moment of resistance of

$$M'_p = M_p - \text{loss}$$

$$= 470.25 - 25.27 = 444.98 \text{ kNm}$$

EXAMPLE 29

The profile in Fig. 50 (a) carries a factored axial thrust of 500 kN. Determine the value of the available plastic moment of resistance.

SOLUTION

Total area of profile is

$$A = 2BT + t_w d$$
$$= 2 \times 200 \times 20 + 10 \times 380$$
$$= 11\,800 \text{ mm}^2$$

now

$$\frac{P}{f_y} = \frac{500 \times 10^3}{250} = 2000 \text{ mm}^2$$

Since this is less than $A_w = 3800 \text{ mm}^2$, use

$$Z_p' = Z_p - \frac{A^2}{4t_w}\left(\frac{P}{P_y}\right)^2,$$

$$P_y = f_y A = 250 \times 11\ 800 \times 10^{-3} = 2950 \text{ kN}$$

Then

$$Z_p' = (1881 \times 10^3) - \frac{11\ 800^2}{4 \times 10}\left(\frac{500}{2950}\right)^2 = 1781 \times 10^3 \text{ mm}^3,$$

giving

$$M_p' = 250 \times 1781 \times 10^3 \times 10^{-6} = 445.3 \text{ kN m}$$

Alternatively Z_p' may be found as follows. Since $P/f_y < A_w = t_w x$

$$x = \frac{2000}{10} = 200 \text{ mm},$$

$$\therefore \text{ loss of plastic modulus} = t_w\frac{x^2}{4}$$

$$= 10 \times \frac{200^2}{4} = 100 \times 10^3 \text{ mm}^3$$

Hence

$$Z_p' = (1881 \times 10^3) - (100 \times 10^3) = 1781 \times 10^3 \text{ mm}^3, \text{ as before.}$$

EXAMPLE 30

By increasing the axial thrust to 1950 kN in the previous problem what effect will this have on the value of Z_p?

SOLUTION

$$\frac{P}{f_y} = \frac{1950 \times 10^3}{250} = 7800 \text{ mm}^3 > A_w$$

Thus zero stress axis will be in flange, hence use

$$Z_p' = \frac{A^2}{4B}\left(1 - \frac{P}{P_y}\right)\left(\frac{2DB - A}{A} + \frac{P}{P_y}\right),$$

$$\frac{P}{P_y} = \frac{1950}{2950} = 0.661,$$

$$\therefore Z_p' = \frac{11\ 800^2}{4 \times 200}(1 - 0.661)\left(\frac{(2 \times 400 \times 200) - 11\ 800}{11\ 800} + 0.661\right),$$

$$= 780 \times 10^3 \text{ mm}^3$$

Alternatively Z_p' may be found as follows. Since $P/f_y > A_w$, whole of web area must be discounted; then area of flanges resisting thrust is

$$7800 - 3800 = 4000 \text{ mm}^2$$

net thickness of each flange available for bending is

$$T - t_t = 20 - \frac{4000}{2 \times 200} = 10 \text{ mm},$$

giving a centroidal distance between these two net flange thicknesses of

$$400 - 10 = 390 \text{ mm},$$

$$\therefore Z_p' = 200 \times 10 \times 390 = 780 \times 10^3 \text{ mm}^3, \text{ as before.}$$

EXAMPLE 31

Design by plastic methods, using grade 43 steel, the two-span continuous beam shown in Fig. 34. The loads are working loads and expressed in kN. Assume uniform profile throughout.

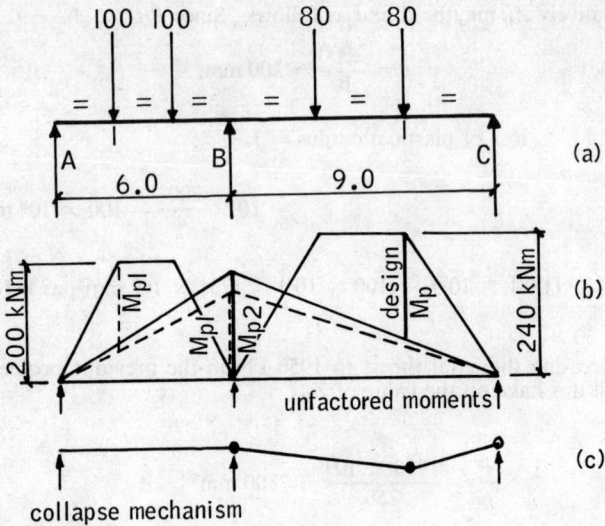

FIG. 51

SOLUTION

The free B.M.s are calculated and drawn to scale (Fig. 51 (b)), omitting the load factor. Superimpose reactant diagrams.

Clearly the maximum net span B.M.s are at points 1 in both bays. In span AB at point 1,

$$200 - \frac{M_{p1}}{3} = M_{p1},$$

giving

$$M_{p1} = \frac{3 \times 200}{4} = 150 \text{ kNm}$$

Due to this value of M_{p1} the corresponding net moment at point 1 in span BC will be

$$M_N = 240 - \frac{150}{3} = 190 > M_{p1}$$

In span BC at point 1,

$$240 - \frac{M_{p2}}{3} = M_{p2}$$

giving

$$M_{p2} = \frac{3 \times 240}{4} = 180 \text{ kNm,}$$

and due to this value of M_p the corresponding net moment at point 1 in span AB will be

$$M_N = 200 - \frac{180}{3} = 140 < M_{p2} \text{ and } M_{p1}$$

From these two cases it may be observed that if the beam is designed for M_{p1} in span AB a larger moment will exist in span BC, however, by designing for M_{p2} in span BC no moment larger than this exists anywhere else. To satisfy yield requirements the design must be based on M_{p2}. Therefore collapse moment is

$$M_p = 1.75 \times 180 = 315 \text{ kN m}$$

modulus required

$$Z_p = \frac{315 \times 10^6}{250} = 1260 \times 10^3 \text{ mm}^3$$

Try $406 \times 178 \times 67$ U.B., $Z_p = 1346 \text{ cm}^3$ $t_w = 8.8$ mm, $d = 380.8$ mm. Shear at support in span BA

$$R_{BA} = 1.75 \left[100 + \frac{180}{6} \right] = 227.5 \text{ kN}$$

giving an average shear stress of

$$q = \frac{227.5 \times 10^3}{8.8 \times 380.8} = 67.9 \text{ N/mm}^2$$

this is less than $250/\sqrt{3} = 144$ N/mm²
    For the web alone

$$Z_p = \frac{8.8 \times 380.8^2}{4} = 319 \times 10^3 \text{ mm}^3$$

and the reduced yield stress is

$$f_y' = \sqrt{(250^2 - 3 \times 67.9^2)} = 220.6 \text{ N/mm}^2$$

giving a loss factor of

$$\frac{f_y - f_y'}{f_y} = \frac{250 - 220.6}{250} = 0.118$$

so that the net available plastic modulus is

$$Z_p' = (1346 - 0.118 \times 319) 10^3 = 1308 \times 10^3 \text{ mm}^3$$

this is greater than the Z_p required, therefore section is satisfactory.
Use 406 × 178 × 67 U.B.

EXAMPLE 32

Design by plastic methods, using grade 43 steel, the two span beam shown in Fig. 52. The loads are unfactored working loads in kN units. Base the design on the smaller span and provide flange plates in the larger span.

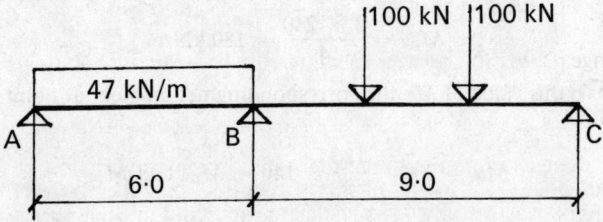

FIG. 52

SOLUTION

Draw the free unfactored B.M.s for both spans and superimpose a reactant B.M. based upon the smaller span (Fig. 53).

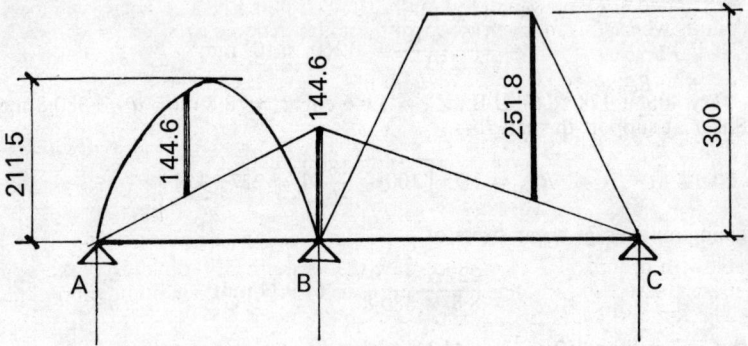

FIG. 53

It may be shown that for a U.D.L. acting on a propped cantilever that the plastic moment has the value $WL^2/11\cdot7$ which will occur at the fixed end and at a distance of $0\cdot414L$ from the simple end. So that for span AB

$$M_p = \frac{47 \times 6^2}{11\cdot7} = 144.6 \text{ kN m}$$

Using a load factor of 1·75 the plastic modulus required is

$$Z_p = \frac{1\cdot75 \times 144\cdot6 \times 10^6}{250} = 1012\cdot2 \times 10^3 \text{ mm}^3$$

The nearest section is a 406 × 178 × 54 U.B., $Z_p = 1048$ cm³, but after allowing for shear the net Z_p would be 940·5 cm³ which is less than that re-

quired. The next section is $406 \times 178 \times 60$ U.B., $Z_p = 1194$ cm³ which, allowing for shear, reduces to 1090·8 cm³ which is adequate.
In span BC at the first point load the net moment is

$$M_p = 300 - \frac{144·6}{3} = 251·8 \text{ kN m}$$

and for a load factor of 1·75 gives

$$Z_p = \frac{1·75 \times 251·8 \times 10^6}{250} = 1762·6 \times 10^3 \text{ mm}^3$$

the flange plates must provide the difference between this value and that of the chosen section, *i.e.*

$$\text{plate } Z_p = (1762·6 = 1194) \, 10^3 = 568·6 \times 10^3 \text{ mm}^3$$

Plastic design techniques allow the addition and subtraction of Z_p values. For the plates, the extra modulus required may be found from $Z_p = BT(d + T)$, where B and T are the plate width and thickness respectively and d is the depth of the U.B.
Then, assuming $T = 12$ mm,

$$568·6 \times 10^3 = 12B \, (406·4 + 12)$$

giving $B = 113.2$ mm
say 115 mm $\times$ 12 mm thick on both flanges
The length of these flange plates must extend a distance not less than that calculated for an unfactored moment of 144·6 kN m. There will be two points where this moment obtains, these are denoted as x_1 and x_2 from C.

$$R_{CB} = 100 - \frac{144·6}{9} = 83·9 \text{ kN}$$

for $x_1 < 3$, $144·6 = 83·9 \, x_1$

giving $x_1 = 1·72$ m
for $x_2 < 9$, $144·6 = 83·9x_2 - 100 \, (x_2 - 3) - 100 \, (x_2 - 6)$
giving $x_2 = 6·51$ m
Now in order to effectively generate the strength of the plates at these positions the starting and finishing points need to be less than x_1 and more than x_2. The force to be carried by the plates is

$$P = 12 \times 115 \times 250 = 345 \text{ kN}$$

Setting the weld yield shear stress at 145 N/mm² and using 6 mm fillet welds, the length of weld required to resist this force at both ends is

$$l = \frac{345 \times 10^3}{0·7 \times 6 \times 145} = 566·5 \text{ mm}$$

since the weld will be returned round the end of the plates by 115 mm, the side lengths will be

$$l_s = \frac{566·5 - 115}{2} = 225·8 \text{ mm, say 230 mm}$$

Therefore the actual starting and finishing positions will be $1·72 - 0·23 = 1·49$ from C and $6·51 + 0·23 = 6·74$ m from C. The plate length will be $6·74 - 1·49 = 5·25$ m.

Another way of finding l_s would be to calculate half the length of the plastic hinge. *See* Fig. 54.

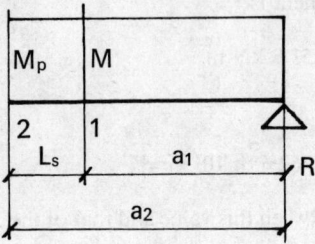

at 2 $\qquad M_p = f_y Z_p = Ra_2$

at 1 $\qquad M = f_y Z = Ra_1$

then $f_y (Z_p - Z) = R(a_2 - a_1)$

where $\quad a_2 - a_1 = l_s$

FIG. 54

Inserting values

$$l_s = \frac{250 \,(1194 - 1058) \,10^3}{1\cdot75 \times 83\cdot9 \times 10^3} = 229 \text{ mm}$$

which gives a reasonably similar answer.

CHAPTER 6

COMPOSITE DESIGN OF BEAMS

IT has been, and still is, common practice to design a steel frame totally independent of the supported floor slabs and subsequent encasement, with the result that stresses attained bear little resemblance with those initially assumed. In short, conventional methods of design underestimated the strength of the structure.

The method of composite construction allows the designer to take into account the simultaneous action between two structural materials which are in close contact and bonded together. A steel beam supporting a reinforced concrete floor slab, and a steel column encased in concrete represent two instances of composite behaviour. B.S. 449 permits the designer to consider the benefits of concrete encasement to both beams and columns, while C.P. 117 deals specifically with the design of beams acting compositely with supported slabs.

No improved method of design is without its disadvantages. Frequently the amount of calculation increases in order to achieve economy, with the result that fabrication methods require updating and construction at site needs closer control to obtain the design standard. Composite construction is no exception.

BASIC DECISIONS

A composite beam may be designed by either elastic or ultimate load methods. Both design approaches require some form of mechanical bond between the beam and slab, and regard must be given to the proposed method of construction, since on it will depend the final stresses. The beam and the slab which it supports cannot be treated in total isolation and therefore the following decisions are pertinent at the commencement of design:

1. Grade of beam steel.
2. Grade of slab concrete.
3. Grade of slab reinforcement.
4. Method of design:

   (a) elastic;
   (b) ultimate load (load factor).

5. Construction method:

   (a) propped against deflection during construction;
   (b) unpropped during construction.

Item 5 (a) in this list is very important and failure to communicate this decision through the contract documents to site control may prove dangerous: propped construction presupposes that the entire loading will be carried by the composite beam on completion. Unpropped construction presupposes that the self-weight of the plain beam and the slab are non-composite, *i.e.* carried by the plain beam alone and that all subsequent loading is carried by the composite beam.

Stress distribution

The methods of stress calculation will be discussed in the following sections. Depending upon the decisions taken above, particularly with regard to items 4 and 5, typical stress distributions will be of the form shown in Figs. 55 (*a–e*), Figs. 55 (*a–c*) showing elastic methods and Figs. 55 (*d–e*) ultimate load methods.

Basic proportions

The slab is first designed in accordance with C.P. 114 or C.P. 110, but the final arrangement of span reinforcement is held over until the finalised composite design. A composite beam is designed as a T beam or L beam as position demands. Effective width of slab to be considered (based upon C.P. 114 rules) in calculation should be the lesser of:

1. For T beams:
    (*a*) twelve times slab thickness + width of beam flange;
    (*b*) one-third of span of beam;
    (*c*) centres between beams.
2. For L beams:
    (*a*) four times slab thickness + width of beam flange;
    (*b*) one-sixth of span of beam;
    (*c*) half the centres between beams.

Calculating the stresses in elastic design or the ultimate moment of resistance in load factor design presupposes a knowledge of the beam size to be employed. As a preliminary "guess" the beam may first be sized taking full restraint, using Chapter 2 methods and Z obtained reduced by 10 per cent to 30 per cent, depending upon the final design approach.

ELASTIC DESIGN METHOD

Calculated working bending stresses in the steel beam must not exceed those given in B.S. 449. Web shear, bearing and buckling are taken on the steel beam only and the stresses should not exceed the relevant B.S. 449 values.

The calculated bending stresses in the concrete resulting from composite behaviour should not exceed $\frac{f_{cu}}{3}$ for nominal mixes and $\frac{f_{cu}}{2.73}$ for designed mixes, where f_{cu} is the 28 day crushing strength (cube).

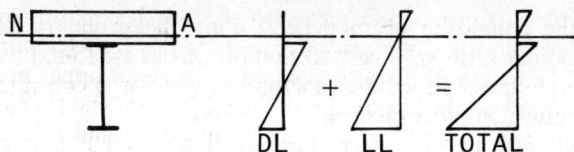

Dead load carried by steel only. Composite neutral axis in slab.

FIG. 55(a)

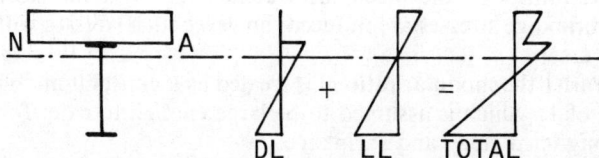

Dead load carried by steel only. Composite neutral axis in steel.

FIG. 55(b)

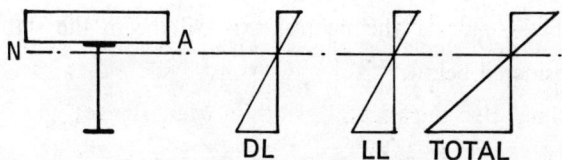

All loads carried by composite beam. The composite neutral axis may be in either the beam or the slab.

FIG. 55(c)

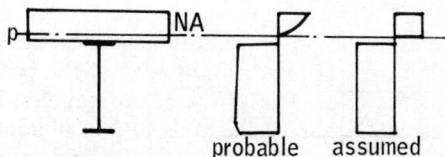

All loads carried by composite beam—neutral axis in slab.

FIG. 55(d)

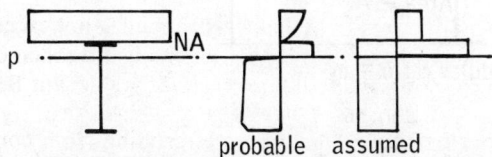

All loads carried by composite beam—neutral axis in beam.

FIG. 55(e)

Properties of the composite beam depend upon the geometry as affected by the modular ratio m. These section properties are found by employing the simple transformed area method in which the concrete is reduced to an equivalent area of steel.

Under the effects of long term permanent loading concrete creeps, *i.e.* the loading strains increase with time, and this is reflected as an apparent reduction in its elastic modulus. Hence the modular ratio $m = E_s/E_c$ increases with time. Similarly concrete shrinks, but this movement is restrained by the mechanical connection with the steel beam, so that shrinkage stresses are induced, however such stresses will be modified by creep.

In C.P. 117 Part 1 the modular ratio m is treated as a constant and is given the value of 15 which is assumed to be large enough to cater for the effects of long term creep and shrinkage.

The neutral axis of a composite beam may be either in the slab or in the steel depending upon the chosen proportions. Referring to Figs. 55 and 56 it may be shown that if $\dfrac{Bd_s^2}{mA_sD} < 1$ the neutral axis will be in the steel beam, and if $\dfrac{Bd_s^2}{mA_sD} > 1$ the neutral axis will be in the slab. Both cases are considered below.

1. *For NA in slab*

Ignore the concrete below the NA. From Fig. 56 by taking moments about the top of the slab,

$$\frac{Bn^2}{2m} + A_s y_s = n\left(\frac{Bn}{m} + A_s\right)$$

and solve for n.

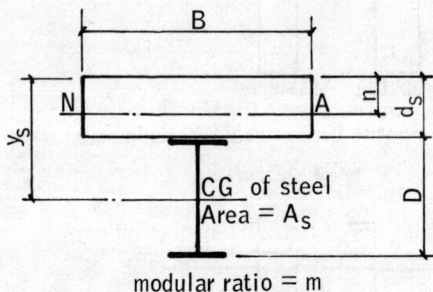

modular ratio = m

FIG. 56

The composite moment of inertia is then

$$I_g = I_s + A_s(y_s - n)^2 + \frac{Bn^3}{3m},$$

giving a section modulus at the top of the slab in concrete units of

$$Z_{cc} = \frac{mI_g}{n}$$

and at the bottom of the steel in tension, in steel units of

$$Z_{st} = \frac{I_g}{D + d_s - n}$$

2. *For NA in beam*

All the concrete may be considered. From Fig. 57, by taking moments about the top of the slab,

$$\frac{Bd_s^2}{2m} + A_s y_s = n\left(\frac{Bd_s}{m} + A_s\right)$$

and solve for n.

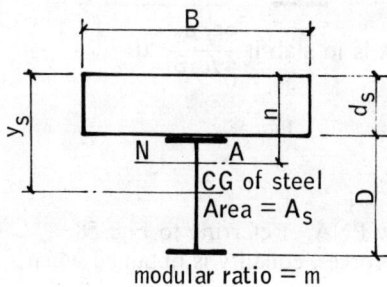

FIG. 57

The composite moment of inertia is then

$$I_g = I_s + A_s(y_s - n)^2 - \frac{Bd_s^3}{12m} + \frac{Bd_s}{m}\left(n - \frac{d_s}{2}\right)^2,$$

giving a section modulus at the top of the slab in concrete units of

$$Z_{cc} = \frac{mI_g}{n}$$

and at the bottom of the steel in tension, in steel units of

$$Z_{st} = \frac{I_g}{D + d_s - n}$$

Actual bending stresses may now be calculated, having regard to whether the section is propped or unpropped, employing the simple theory of bending. The bending moment to be resisted is found in the usual manner.

ULTIMATE LOAD METHOD

Stresses are not calculated by this method. They are assumed to be the yield stress of the steel and four ninths of the 28-day strength of the concrete. The position of the neutral axis, or more correctly the equal force axis, is dependent upon the full "plastic" strength of the composite section. It is assumed that the steel is stressed uniformly to its yield strength f_y and the concrete stressed uniformly to $4f_{cu}/9$.

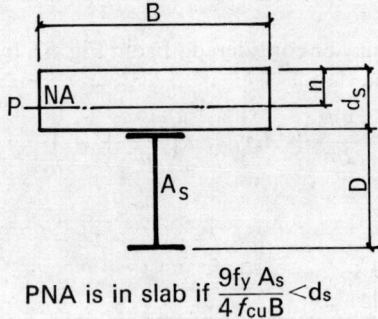

PNA is in slab if $\dfrac{9f_y A_s}{4f_{cu}B} < d_s$

Fig. 58

1. *For PNA in slab*

Ignore concrete below PNA. Referring to Fig. 58, since there are no axial forces, equality is obtained when

$$\tfrac{4}{9}f_{cu}Bn = f_y A_s,$$

giving

$$n = \frac{9f_y A_s}{4f_{cu}B}$$

and taking moments about the centroid of the steel the ultimate moment of resistance is

$$M_u = \tfrac{4}{9}f_{cu}Bn\left(\frac{D}{2} + d_s - \frac{n}{2}\right)$$

or about the centroid of the concrete in compression,

$$M_u = f_y A_s\left(\frac{D}{2} + d_s - \frac{n}{2}\right)$$

Both these expressions will be equal.

2. *For PNA in steel*

All the concrete may be considered. Referring to Fig. 59,

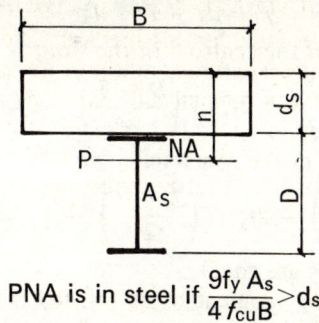

PNA is in steel if $\dfrac{9f_y A_s}{4 f_{cu}B} > d_s$

FIG. 59

since there are no axial forces, then equality is obtained when

$$\tfrac{4}{9}f_{cu}Bd_s + A_{sc}f_y = A_{st}f_y;$$

this expression involves two unknown quantities, A_{sc} and A_{st}. Putting $A_{st} = A_s - A_{sc}$ obtains

$$\tfrac{4}{9}f_{cu}Bd_s + 2A_{sc}f_y = A_sf_y$$

which reduces the number of unknowns to one. However, the distance to the PNA is still undetermined. Let b_f be the width of the steel flange and T its thickness then if

$$t_x = \frac{A_sf_y - \tfrac{4}{9}f_{cu}Bd_s}{2f_yb_f} < T$$

the PNA is in the steel flange, and

$$n = d_s + t_x$$

Taking moments about the centroid of the steel,

$$M_u = \tfrac{4}{9}f_{cu}Bd_s\left(\frac{D + d_s}{2}\right) + 2f_yb_ft_x\left(\frac{D - t_x}{2}\right),$$

or about the centroid of the concrete,

$$M_u = f_yA_s\left(\frac{D + d_s}{2}\right) - 2f_yb_ft_x\left(\frac{d_s + t_x}{2}\right)$$

Both these expressions will be equal.

If $t_x > T$ the PNA is in the web. Let t_w be the web thickness and d_x be the depth of the web in compression measured from the underside of the flange then

$$d_x = \frac{A_sf_y - \tfrac{4}{9}f_{cu}Bd_s - 2b_fTf_y}{2t_wf_y}$$

giving $$n = d_s + T + d_x$$

Taking moments about the centroid of the steel,

$$M_u = \tfrac{4}{9}f_{cu}Bd_s\left(\frac{D + d_s}{2}\right) + 2f_y b_f T\left(\frac{D - T}{2}\right) + 2f_y t_w d_x\left(\frac{D}{2} - T - \frac{d_x}{2}\right)$$

or about the centroid of the concrete,

$$M_u = f_y A_s\left(\frac{D + d_s}{2}\right) - 2f_y b_f T\left(\frac{d_s + T}{2}\right) - 2f_y t_w d_x\left(\frac{d_s}{2} + T + \frac{d_x}{2}\right)$$

Both these expressions are equal.

The external bending moment to be resisted is found by the usual methods and multiplied by a load factor taken as 1·75.

DETAILED DESIGN PROCEDURES

Provision of shear connectors

Shear connectors are welded to the top flange of the beam and embedded into the concrete slab during casting. They provide the mechanical bond to ensure composite behaviour which is obtained by preventing slipping between the two elements. These connectors must also provide against the tendency of the slab to lift away from the beam when under load.

Various types of connectors are illustrated in the code, and ultimate load values are given in relation to the grade of concrete employed for the slab. The number of shear connectors required between the position of zero and maximum factored bending moment is

$$N = \frac{\text{ultimate load of concrete in compression}}{\text{ultimate load of connector}}$$

and the quantity obtained may be evenly spaced between these two positions. The spacing, however, should not exceed four times the slab thickness or 600 mm, whichever is the smaller.

When a composite beam is designed against heavy concentrated loads the number of shear connectors found above should be shared between the areas of the shear force diagram measured between the positions of discontinuity. To guard against uplift, connectors should be at least 50 mm high and project at least 25 mm into the compression zone of the concrete.

Shear connectors need not be provided in elastic composite design if the steel beam is solidly encased in concrete in accordance with B.S. 449 and the top flange of the beam is at least 50 mm above the soffit of the slab.

Horizontal shear in slab

The slab must have sufficient shear capacity to resist the splitting and punching action due to the force introduced by the connectors. If N_c is the number of connectors at a cross section, P_c the connector force at ultimate load is kilonewtons and s their spacing in millimetres then the actual force per millimetre run is

$$Q = \frac{N_c P_c \times 10^3}{s}$$

This force must not exceed

$$Q_1 = 0 \cdot 23 L_s \sqrt{f_{cu}} + A_t f_y n,$$

or

$$Q_2 = 0 \cdot 62 L_s \sqrt{f_{cu}}$$

where A_t = area of bottom reinforcing steel in slab per millimetre run of beam;

f_y = yield stress of reinforcing steel;

L_s = length of shear surface at connectors, limited to a maximum of $2d_s$ for T beams and d_s for L beams;

n = number of times a reinforcing bar is intersected by the shear surface, twice for T beams, once for L beams.

Deflection

The deflection of a composite beam should be checked elastically at working load conditions and should not exceed the B.S. 449 limits. m should be taken as 15 for imposed loads and 30 for dead loads.

EXAMPLE 33

A floor consists of a series of beams $8 \cdot 0$ m span and $4 \cdot 0$ m apart, supporting a reinforced concrete slab 120 mm thick which is reinforced with 10 mm diameter mild steel bars at 150 mm centres. If the superimposed load is $2 \cdot 5$ kN/m² and the slab finishes are $1 \cdot 0$ kN/m², design the supporting beams for: (1) elastic unpropped, (2) elastic propped, (3) ultimate load propped.

Assume the beams are grade 43 steel and the concrete strength 21 N/mm² at 28 days. Take concrete at 24 kN/m³.

SOLUTION

1. Elastic unpropped

Loading	Slab	$\dfrac{4 \times 120 \times 24}{1000}$	$= 11 \cdot 52$ kN/m
	Finishes	$4 \times 1 \cdot 0$	$= 4 \cdot 00$
	Impose	$4 \times 2 \cdot 5$	$= 10 \cdot 00$
	S.W. beam		$0 \cdot 6$

During construction the dead load is due to the slab and the beam self-weight. The finishes are added afterwards and induce composite action together with the imposed load. Dead load bending moment:

$$M_d = \frac{(11 \cdot 52 + 0 \cdot 6) \times 8^2}{8} = 96 \cdot 96 \text{ kNm}$$

Imposed load bending moment:

$$M_i = \frac{(10 + 4) \times 8^2}{8} = 112 \cdot 0 \text{ kNm}$$

Try a $406 \times 178 \times 60$ U.B., $I_{xx} = 21\,520 \text{ cm}^4$, $Z_{xx} = 1059 \text{ cm}^3$, $A = 76 \cdot 1 \text{ cm}^2$. Effective slab width:

$$B = \frac{8000}{3} = 2666 \text{ mm}$$

$$\begin{aligned} \text{or} \quad &= 4000 \\ \text{or} \quad &= 12 \times 120 + 178 = 1618 \text{ (this governs)} \end{aligned}$$

use $B = 162$ cm.

The position of the composite neutral axis will be in the steel beam if Bd_s^2/mA_sD is less than unity, and in the slab if greater (*see* Fig. 60).

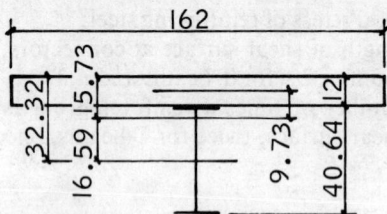

All dimensions in centimetres

FIG. 60

$$\therefore \text{ neutral axis factor} = \frac{162 \times 12^2}{15 \times 76 \cdot 1 \times 40 \cdot 6} = 0 \cdot 5 < 1$$

Since NA is in steel then

$$\frac{162 \times 12 \times 6}{15} + (76 \cdot 1 \times 32 \cdot 32) = n \left[\frac{162 \times 12}{15} + 76 \cdot 1 \right]$$

giving $n = 15 \cdot 73$ cm.

$$I_g = 21\,520 + (76 \cdot 1 \times 16 \cdot 59^2) + \frac{162 \times 12^3}{15 \times 12} + \frac{162 \times 12 \times 9 \cdot 73^2}{15}$$

$$= 56\,364 \text{ cm}^4$$

For concrete

$$Z = \frac{56\,364 \times 15}{15 \cdot 73} = 53\,730 \text{ cm}^3$$

for steel,

$$Z = \frac{56\,364}{36 \cdot 91} = 1528 \text{ cm}^3$$

The dead load stresses on the steel beam only are

$$f_{bc} = f_{bt} = \frac{96 \cdot 96 \times 10^6}{1059 \times 10^3} = 91 \cdot 5 \text{ N/mm}^2,$$

and the composite stresses are, for concrete,

$$f_{bc} = \frac{112 \times 10^6}{53\ 730 \times 10^3} = 2 \cdot 09 \text{ N/mm}^2$$

for steel

$$f_{bt} = \frac{112 \times 10^6}{1528 \times 10^3} = 73 \cdot 4 \text{ N/mm}^2$$

Total steel stress (tension) = $164 \cdot 9$ N/mm^2 < p_{bc} = 165. The composite section selected is adequate for bending.

Shear connections are required to give composite action against the imposed load bending moment. The number required is found from ultimate load methods.

$$A_s f_y = 250 \times 7610 \times 10^{-3} = 1902 \text{ kN,}$$

$$\tfrac{4}{9} f_{cu} B d_s = \tfrac{4}{9} \times 21 \times 1620 \times 120 \times 10^{-3} = 1813 \text{ kN,}$$

therefore PNA is in the steel, then

$$t_t = \frac{(1902 - 1813)10^3}{2 \times 250 \times 178} = 1 \text{ mm,}$$

$$M_u = \left[(1813 \times 263 \cdot 2) + \left(\frac{2 \times 250}{1000} \times 178 \times 1 \times 202 \cdot 7 \right) \right] 10^{-3} = 495 \text{ kNm}$$

Total factored bending moment carried is

$$M = 1 \cdot 75(96 \cdot 96 + 112) = 365 \cdot 8 \text{ kNm}$$

Then force to be resisted by shear connectors is

$$F_s = \frac{1813 \times 366 \cdot 3}{495} = 1340 \text{ kN}$$

Using headed stud connectors 75 mm $\times$ 19 mm (3 in $\times$ $\tfrac{3}{4}$ in), height $\times$ diameter, $P_c = 67$ kN then

$$N = \frac{1342}{67} = 20$$

or 40 connectors in total span. This gives a spacing of about 200 mm which is less than $4 \times d_s$ or 600 and is satisfactory.

The actual force on each connector is $P_c = 67$ kN,

then the shear force in the slab along the connector line is

$$Q = \frac{67 \times 10^3}{190} = 353 \text{ N/mm}$$

Now

$$Q_1 = 0 \cdot 23 L_s \sqrt{f_{cu}} + A_t f_y n,$$

where

$$L_s = (2 \times 75) + 30 = 180 < 2d_s;$$

$$A_t = \frac{10^2 \times \pi}{4} \times \frac{1}{150} = 0.52 \text{ mm}^2/\text{mm};$$

$$f_y = 250 \text{ N/mm}^2, \text{ say};$$
$$n = 2.$$

Then

$$Q_1 = (0.23 \times 180\sqrt{21}) + (0.52 \times 250 \times 2)$$

$$= 450 \text{ N/mm} > Q$$

also

$$Q_2 = 0.62 L_s \sqrt{f_{cu}}$$

$$= 0.62 \times 180\sqrt{21}$$

$$= 512 \text{ N/mm} > Q$$

Since Q does not exceed either Q_1 or Q_2 the section selected is satisfactory for shear requirements.

Vertical shear, bearing and buckling should be checked as for a simple non-composite beam. These will be found to be adequate.

Deflection. The actual L/D ratio is

$$\frac{L}{D} = \frac{8000}{120 + 406.4} = 15.2$$

This ratio is low and deflection will not be a criterion. However, if L/D is greater than 25 deflection needs investigating.

2. Elastic propped

The total load is now to be carried by the composite beam. Since the only change in the loading will be due to the self-weight of the steel beam, which is very small, no loss of accuracy will obtain if the same total loading is assumed here. Total bending moment on composite beams is

$$M = M_d + M_i$$
$$96.96 + 112 = 209 \text{ kN m, say}.$$

Try $406 \times 178 \times 54$ U.B. $I_{xx} = 18\,576 \text{ cm}^4$, $A = 68.3 \text{ cm}^2$. Take effective slab width as before, and the neutral axis factor will be less than unity (*see* Fig. 61), then

$$\frac{162 \times 12 \times 6}{15} + (68.3 \times 32.13) = n\left[\frac{162 \times 12}{15} + 68.3\right],$$

giving $n = 15.02$ cm.

$$I_g = 18\,576 + (68.3 \times 17.11^2) + \frac{162 \times 12^3}{15 \times 12} + \frac{162 \times 12 \times 9.02^2}{15}$$

$$= 50\,761 \text{ cm}^4$$

For concrete,

$$Z = \frac{50\,761 \times 15}{15.02} = 50\,760 \text{ cm}^3,$$

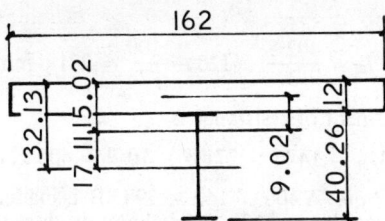

All dimensions in centimetres

FIG. 61

for steel

$$Z = \frac{50\,761}{37 \cdot 24} = 1350 \text{ cm}^3$$

Thus the composite stresses are, for concrete,

$$f_{bc} = \frac{209 \times 10^6}{50\,760 \times 10^3} = 4 \cdot 12 \text{ N/mm}^2$$

for steel,

$$f_{bt} = \frac{209 \times 10^6}{1350 \times 10^3} = 154 \cdot 9 \text{ N/mm}^2$$

Both stresses are less than permissible values: therefore section is adequate for bending. The number of shear connectors required is determined as in the previous example.

3. Ultimate load propped

The loading is as before and there is no point in modifying the beam self-weight. The ultimate bending moment to be accommodated is

$$M_u = 1 \cdot 75(M_d + M_i)$$
$$= 1 \cdot 75 \times 209 = 366 \text{ kNm}$$

Try $406 \times 140 \times 46$ U.B., $A = 58 \cdot 9$ cm^2. The effective slab width will now be

$$B = (12 \times 120) + 140 = 1580 \text{ mm}$$

Now the PNA will be in the slab if

$$\frac{9 \times f_y A_s}{4 \times f_{cu} B} < d_s$$

then

$$\frac{9 \times 250 \times 5890}{4 \times 21 \times 1580} = 100 < d_s = 120$$

Ultimate load on concrete is

$$F_{cc} = \tfrac{4}{9} \times 21 \times 1580 \times 100 \times 10^{-3} = 1474 \text{ kN},$$

ultimate load on steel is

$$F_{st} = 250 \times 5890 \times 10^{-3} = 1474 \text{ kN}$$

The lever arm is

$$l_a = \frac{402 \cdot 3}{2} + 120 - \frac{100}{2} = 271 \cdot 6 \text{ mm}$$

Hence ultimate moment of resistance is

$$M_u = 1476 \times 271 \cdot 6 \times 10^{-3} = 400 \text{ kNm}$$

which will suffice. *Note:* A $406 \times 140 \times 39$ U.B. is inadequate, and shallower sections having a mass between 39 and 46 kg/m produce too small a lever arm to give the required moment.

Shear connectors. Using headed stud connectors 75 mm height $\times$ 19 mm diameter (3 in $\times \frac{3}{4}$ in), $P_c = 67$ kN, then

$$N = \frac{1474}{67} = 22$$

that is, 44 connectors on the total span, giving a spacing of 180 mm.

The shear force along the connector line is

$$Q = \frac{67 \times 10^3}{180} = 372 \text{ N/mm},$$

and

$$\left.\begin{array}{l} Q_1 = 450 \text{ N/mm} \\ Q_2 = 512 \text{ N/mm} \end{array}\right\} \text{as in the first example.}$$

The proposed section must be checked under working load conditions to ensure that the concrete stress does not exceed $0 \cdot 33 \times f_{cu} = 7$ N/mm² and the steel stress does not exceed $0 \cdot 9 f_y = 225$ N/mm². The method of calculating the properties and stresses is as for the previous example. In fact these requirements will be seen to be satisfied.

EXAMPLE 34

The beam in Fig. 62 has the loading proportioned to give an identical beam to that obtained in the previous example. Determine the disposition of the shear connectors if 22 are required in half the span.

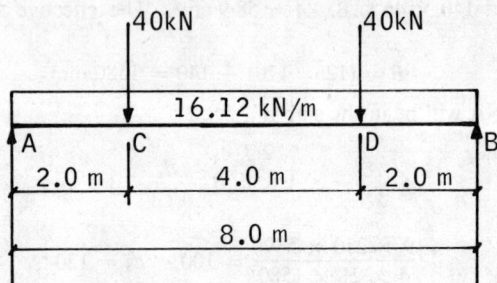

Fig. 62

SOLUTION

$$\text{At A the shear} = (4 \times 16 \cdot 12) + 40 = 104 \cdot 48 \text{ kN}$$
$$\text{At the left of C the shear} = 104 \cdot 48 - (2 \times 16 \cdot 12) = 72 \cdot 24 \text{ kN}$$

At the right of C the shear $= 72 \cdot 24 - 40 = 32 \cdot 24$ kN
At the centre the shear $= 0$

Between A and C the area of the shear force diagram is

$$A_1 = \left(\frac{104 \cdot 48 + 72 \cdot 24}{2}\right) \times 2 = 176 \cdot 72$$

Between C and the span centre,

$$A_2 = \frac{32 \cdot 24 \times 2}{2} = 32 \cdot 24$$

giving a total area of $176 \cdot 72 + 32 \cdot 24 = 208 \cdot 96$.
The number of shear connectors required between A and C is

$$N_1 = \frac{A_1}{A_1 + A_2} N = \frac{176 \cdot 72 \times 22}{208 \cdot 96} = 18 \cdot 6, \text{ say } 19,$$

and the remainder are placed between C and the span centre, that is three.

However the spacing is nevertheless restricted to $4d_s = 480$ or 600 mm, whichever is the least, so the number of connectors must be increased to four.

In the final layout the connector spacing will be 100 mm between A and C and almost 445 mm between C and D. Of course the distance B to D will be as A to C.

INDEX

steelwork connections

Methods.

Examples of Use.

Method of detailing

12/2/87

$$\underline{\qquad 31 \qquad} - \overline{|\,19\,|}$$

50

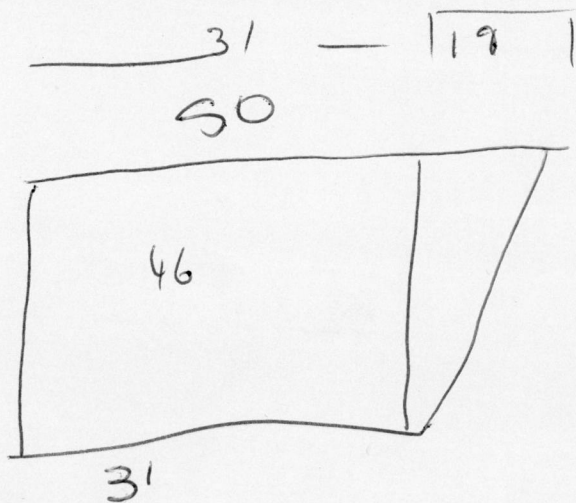

46

31

12,834,

16,767 m^3